The Ontogeny of Information

Science and Cultural Theory

A Series Edited by Barbara Herrnstein Smith

and E. Roy Weintraub

The Ontogeny

of Information

Developmental Systems and Evolution

Second Edition, Revised and Expanded

Susan Oyama

Foreword by Richard C. Lewontin

DUKE UNIVERSITY PRESS   2000

Second edition, revised and expanded

2nd printing, 2002

© 2000 by Duke University Press

Originally published in 1985 by Cambridge University Press

All rights reserved

Printed in the United States of America on acid-free paper ∞

Designed by C. H. Westmoreland

Typeset in Times Roman with Din Neuzeit display

by Tseng Information Systems, Inc.

Library of Congress Cataloging-in-Publication Data

appear on the last printed page of this book.

Contents

Foreword

The first edition of *The Ontogeny of Information* (1985) was instigated by a growing impoverishment of explanation and understanding of how organisms come to be. As a consequence of a long history of discovery in genetics and especially in molecular genetics, together with a virtually complete stagnation of the classical science of experimental embryology, explanations of how organisms grow and differentiate to produce their characteristic forms and properties had become framed entirely in terms of the "turning on" and "turning off" of genes. Already in 1982, at the Cambridge University symposium commemorating the one hundredth anniversary of Darwin's death, Sydney Brenner in his keynote address had declared that given the complete DNA sequence of an organism and a large enough computer it would soon be possible to compute the organism. By 1985, the genetic boa constrictor had totally enfolded embryology in its helical coils. Now, nearly fifteen years later, the helpless victim has disappeared down the molecular maw and is slowly being digested. In the end, of course, the result will be what is to be expected from such a process: a slight enlargement of the body of the consumer and the production of a large amount of fecal matter.

As Susan Oyama explains in *The Ontogeny of Information,* there are two difficulties with the current trend of explanation of development. The first is that the wrong question is being asked; the second is that, even to the extent that we are concerned with the answer to that question, the wrong answer is being given. When the wrong question is being asked, it usually turns out to be because the right question is too difficult. Scientists ask questions they can answer. That is, it is often the case that the operations of a science are not a consequence of the problematic of that science, but that the problematic is induced by the available means. The geneticist Al Hershey used to say that heaven is finding an experiment that works, and then doing it over and over and over. Sequencing DNA is easy; any damn fool with a PCR machine can do it. It is rather more difficult to make genetic constructs of just the right sort to alter or knock out the reading of particular genes, but a competent *Drosophila* geneticist can usually manage it. On the other hand, no one seems to know even how to *ask* an experimental question about the actual coming together of molecules to make cell organelles, or why my rear appendages look like legs and feet while my front appendages are arms and hands, or why a particular microanatomy of the central nervous system allows the reader to understand this page.

The reason that the wrong answer is being given, even to the questions we know how to ask, is that some forms of thinking we depend on lead us astray. First, we do not know how to dispense with metaphorical language and are often victimized by properties of the metaphor. Physicists can speak of the "billiard ball" model of colliding molecules without supposing that molecules are made of plastic, or that they are red and white, or that they make a clicking sound when they hit each other. But when biologists speak of genes as "computer programs," they erroneously suppose that all the organism's attributes are prefigured in its genes and all that is required for a fixed output is for the *enter* key to be pressed.

Second, unlike physicists, who have learned to be rather more sophisticated, biologists, who imagine that they are following the example of physics, have a simplistic notion of causes and effects. If A and B are causally connected, they think, then either A is the cause and B the effect or vice versa. But no physicist would try to identify the separate cause and effect in the coupled electromagnetic equations. The magnetic field does not cause the electric field any more than the electric field causes the magnetic. There is simply a pair of coupled differential equations describing the electromagnetic phenomena that are two aspects of the same process. Nor is it the case, as biologists usually believe, that equal causal forces have equal effects. In a dynamical process in time it is also necessary to take account of initial conditions. If I drive eighty miles west and twenty-five miles north, it is impossible for you to know that I will arrive in Brattleboro, Vermont, unless you know that I started in Boston. Nor can physical processes be counted on to behave in a determinate way when they are based on a very small number of reactive molecules irregularly distributed in space, as is the case for most of the important reactants within cells.

Throughout the history of modern biology there has been a confusion between two basic questions about organisms: the problem of the origin of differences and the problem of the origin of state. At first sight these seem to be the same question, and taken in the right direction, they are. After all, if we could explain why each particular organism has its particular form, then we would have explained, *pari passu,* the differences between them. But the reverse is not true. A sufficient explanation of why two things are different may leave out virtually everything needed to explain their nature. Suppose that on the railway line between Boston and New York there is a switch that, when set to the right, sends the train on a direct southwesterly route between the cities, but, when set to the left, shunts the train on a circuitous path westward and then southward along the Hudson River. In response to the question, "Why did it take you eight hours to get to New York from Boston instead of the usual five?" I might quite properly answer that the switch was set to the left, but that tells nothing about the immense complexity of mechanical and social forces involved in the phenomenon of train travel. Indeed, the scheme of explanation of form offered by modern developmental genetics is that of a signaling network among genes that resembles nothing so much as the illuminated map on the wall of a railroad control room with red and green lights twinkling on and off as the switches are reset. The temptation to substitute the explanation of differences for an explanation of outcomes, substituting a map of the switches for an articulated physical and social explanation of the movements of trains, has plagued biology at least since the middle of the nineteenth century.

The experiments of Mendel on variation in peas were not motivated simply by idle intellectual curiosity; they were part of a research program motivated by fruit breeding in Moravia, a program to discover the quantitative laws of *variation* in plants. Mendel's paper was entitled "Experiments on Plant Hybrids," and he tells us in the "Introductory Remarks" that among prior investigations by others, "not one has been carried out to such an extent and in such a way as to make it possible to determine the number of different forms under which the offspring of hybrids appear, or to arrange these forms with certainty according to their separate generations, or definitely to ascertain their statistical relations." Not a word here about the origin or inheritance of form itself. These were not experiments meant to explain why some flowers are red and others white, why some plants are tall and others short. He was apparently concerned only with the "statistical relations" of differences. It was Mendel's single-minded concentration on variation, on the statistical regularities of *differences* arising in the offspring of hybrids, that was his epistemological break from the past which made possible the science of genetics. Yet, at the very moment of his radical move from studying similarities to studying differences, Mendel himself slipped over into an explanation of the origin of form itself by postulating that the red- and white-flowered plants had mysterious "factors" *for* redness and whiteness. Moreover, the factor for redness had a power that enabled it to dominate the factor for whiteness when the two were present in the same plant. The language of the original problematic and of the description of the results, a language entirely of differences, becomes, in the explanatory hypothesis, a language of the origin of form. So at its very birth, the study of the laws of variation in heredity prepared to be the study of the heredity of type. Genetics prepared to swallow and digest the problematic of embryology. The subsequent history of genetics is in large part the story of this ingestion.

During the first half of the present century the major goal of genetics, in addition to the elaboration of the rules of assortment of genetic differences in heredity, was to uncover the material basis of which the rules are the quantitative manifestation. But this search for a material basis has been a kind of reinforcing practice for the theoretical view that the explanation of differences is the same as the explanation of outcomes. This is seen even in the naming of genes. Geneticists speak casually of the "gene for white eyes," but, of course, there is no such gene. There is a variety of genes whose reading by the cell is proximally involved in the production of eye pigment and its deposition in the eye cells. One of these, the so-called white locus, can be so altered from its usual form by mutations that no eye pigment can be deposited. This gene is neither the gene "for" white eyes nor the gene "for" red eyes, and geneticists know that very well, yet the constant repetition of the language of genes "for" white eyes or "for" vestigial wings has a powerful effect on the conceptualization of the developmental process. Hobbes's witty remark that "words are wise men's counters, they do but reckon with them, but they are the money of fools" fails to take account of the inevitable fetishism that arises from the constant use of counters.

Geneticists' concentration on the origin of differences, and their confusion of this question with the processes leading to an organism's state, arises from a deep structural limitation of the experimental investigation in biology. Organisms, unlike simpler physical systems such as stars and their planets, are medium sized and internally functionally heterogeneous. As a result, they are the nexus of a very large number of weakly determining causal pathways. To determine the motion of a planet I need know only its mass, distance from the sun, angular velocity, and the path of a couple of nearby bodies. For the development and behavior of an organism I need to know vastly more. But how are we to investigate the network of weakly determining causal pathways except by experimentally (i.e., unnaturally) holding constant as many variables as possible and making (unnaturally) large perturbations in one or two other variables — large enough, that is, to see an unambiguous effect. When we do this, however, we have a problem of extrapolation. Would the effect of the perturbation be the same if we had other values of the controlled variables, or, worse, would the effect of small perturbations be simply scaled-down effects of the unnaturally large ones, or are there threshold effects?

These issues translate, for the geneticist interested in development, into holding the environment and as many of the genes as possible constant while introducing a major perturbation of some particular gene. That major perturbation is usually a so-called knockout, a change in the gene so drastic that the cell is unable to read it in a way that is sensibly interpretable by the cell machinery. But if I successfully perform knockout experiments on every gene that can be seen in such experiments to have an effect on, say, wing shape, have I learned the "causes" of wing shape? Have I even learned what causes the wing shape of one species or individual to differ from that of another? After all, two species of *Drosophila* presumably have the same set of relevant loci.

The problem with environmental perturbation is worse. The shapes of wings of *Drosophila* can be altered by changing the temperature of development while holding the genotype constant. In the absence of a completely articulated story of wing formation, the observations of different shapes in different environments remain nothing but raw observations. They cannot be converted into functional explanations. Finally, what are we to do when both genotype and environment vary, as they in fact do in nonexperimental situations? It is an empirical fact that there is no way to combine information from genetic variation experiments in a constant environment with environmental variation experiments with a constant genotype, to predict the outcome of developmental processes when both genotype and environment may vary.

The actual path taken by geneticists is precisely to study only those gene alterations that have major effects on the development of their test organisms while growing those organisms in controlled environments that maximize the developmental effect of the gene alterations. So, only a small proportion of all single gene mutations known by *Drosophila* geneticists are so-called Rank I mutants, that is, mutants in which every individual carrying the genotype is unambiguously different from the nonmutants. And even these mutations are often unambiguous in their effect only at a particular temperature or humidity and only if the background genotype is selected to maximize the mutational effect. Yet these extraordinary situations are the only tools available to the developmental geneticist who is supposed to be interested in the production of form and function. One asks the questions that one knows how to ask.

When we turn from questions to answers we immediately confront the tyranny of metaphors. Indeed, the biological problem itself is described by a metaphor, the problem of "development." The word "development" (*desarrollo* in Spanish, *Entwicklung* in German) means literally an unfolding or unrolling of what is already present, a making manifest of what is already latent, as when we "develop" an exposed film to make manifest the latent image. But there is more than mere etymology here. As Oyama points out, in the struggle between preformationists and epigenicists it is really the preformationists who won, although we do not believe that there is a tiny man in each sperm cell. Instead, it is claimed that there is the "blueprint" or "computer program" or "information" necessary to specify the organism already present in the fertilized egg. But what important difference is there except in mechanical details between a preformed individual and all the information necessary to specify that individual? The metaphor of "development" rules our problematic.

Virtually all the language of genetics describes the role of genes in the formation of organisms. Genes are said to be "self-replicating," they engage in "gene action," they "make" proteins, they are "turned on" or "turned off" by "regulatory" DNA. But none of this is true. Genes are certainly not "self-replicating." New DNA strands are synthesized by a complex cell machinery (another metaphor!) consisting of proteins that use the old DNA strands as templates. We do not speak of manuscripts "self-replicating" in a photocopy machine. Genes "do" nothing, they "make" nothing, they cannot be "turned on" or "turned off" like a light or a water tap, because no energy or material is flowing through them. DNA is among the most inert and nonreactive of organic molecules; that is why stretches of DNA can be recovered intact from fossils long after all the proteins have been lost.

We cannot deny the importance of genes in development. In the absence of DNA there would be no development at all because there would be nothing for the cell machinery to read; indeed, the cell machinery would soon disappear. Moreover, the amino acid sequence of the proteins produced by the cell machinery depends on what sequence of DNA the cell reads. In that restricted sense the DNA can be said to have "information" relevant to the formation of the organism. Yet the manifestation of that sequence as information about the organism occurs only by means of the activity of the cell. For example, we know that a given amino acid sequence does not have a unique folding pattern, but may reach different local free energy configurations depending on the cell milieu in which it folds. More radically, we now know that a particular stretch of DNA has alternative reading blocks, depending on the reading machinery of the cell, so that the same sequence of DNA may contain several different overlapping or included "genes." What is a "gene" is determined in some as yet mysterious way by the cell machinery. So DNA becomes "information" about the organism only in the actual process of cell function. These are concrete examples of the claim about "the ontogeny of information" that is the central point of Oyama's book.

Finally, we need to reconsider the way in which we understand causes. It is in this respect that *The Ontogeny of Information* is more radical than the notion of "interaction" may imply. Oyama is entirely correct that in the production of an organism one cannot regard the internal and the external, genes and environment, as alternative causes. It is bad biology to describe some aspects of the organism as resulting from environmental influence and some as the result of genetic effects. It is, of course, true that the *differences* between organisms can, in some special cases, be ascribed entirely to genetic or environmental *differences*. The children of African slaves, born in North America, had the same melanin pigmentation as their parents, and the offspring of Belgian colonial administrators in the Congo were unmistakably Caucasian. On the other hand, the experience of immigration into North America has shown that differences in the phonemic structure of speech depend on differences in the environment and not on differences in genotype. The children of Eastern European, Italian, and Irish immigrants in Boston all speak indistinguishable American English. But these examples are rare for variation within species; for the most part, differences between individuals are a consequence of both genetic and environmental differences.

Nor can we partition the variation quantitatively, ascribing some fraction of the variation to genetic differences and the remainder to environmental variation. Every organism is the unique consequence of the reading of its DNA in some temporal sequence of environments and subject to random cellular events that arise because of the very small number of molecules of each kind in each cell. While we may calculate statistically an average difference between the carriers of one genotype and another, such average differences are abstract constructs and must not be reified with separable concrete effects of genes in isolation from the environment in which the genes are read. In the first edition of *The Ontogeny of Information* Oyama characterized her construal of the causal relation between genes and environment as interactionist. That is, each unique combination of genes and environment produces a unique and a priori unpredictable outcome of development. The usual interactionist view is that there are separable genetic and environmental *causes,* but the effects of these causes acting in combination are unique to the particular combination. But this claim of the ontologically independent status of the causes as causes, aside from their interaction in the effects produced, contradicts Oyama's central analysis of the ontogeny of information. There are no "gene actions" outside environments, and no "environmental actions" can occur in the absence of genes. The very status of environment as a contributing cause to the nature of an organism depends on the existence of a developing organism. Without organisms there may be a physical world, but there are no *environments.* In like manner no *organisms* exist in the abstract without environments, although there may be naked DNA molecules lying in the dust. Organisms are the nexus of external circumstances and DNA molecules that make these physical circumstances into causes of development in the first place. They become *causes* only at their nexus, and they cannot exist as *causes* except in their simultaneous action. That is the essence of Oyama's claim that information comes into existence only in the process of ontogeny. It is this claim about causes that Oyama, in this new edition, calls "contructivist interactionism," but that I would characterize as *dialectical* in order to emphasize its radical departure from conventional notions of interaction.

The need for a careful reexamination of explanations in developmental biology becomes more pressing as the appearance of success of reductive genetic explanation grows. Hence the occasion for a reissue of *The Ontogeny of Information*. Susan Oyama's dissection of the metaphor of information is critical. It is impossible to carry out scientific explanation without metaphors. Indeed, we can hardly speak without them. The most we can demand is that we be conscious of the metaphorical content of our words and not be carried away when we write of the "cell machinery" which "reads" the DNA during the process of "development." No metaphors are totally benign and without dangers. As Norbert Weiner observed, "The price of metaphor is eternal vigilance."

R. C. Lewontin

Preface to Second Edition

The aim of this reissue is to make *The Ontogeny of Information* available again, roughly a decade and a half after its first appearance. Changes to the text, while many, were mostly to increase clarity or concision; the arguments have been left intact. Because of the continuing terminological disorder surrounding the notion of interaction, I have used the term "constructivist interactionism" throughout to refer to my own views; other varieties are always in quotation marks. In addition, because "environmentalism" has come to refer almost exclusively to issues of ecological conservation, I have used "environmental determinism" to indicate biological determinism's opposite pole in the classic duality. It signals an emphasis on environmental influences in development, if not the absolutism implied by "determinism." Although I did not attempt to update the book, references to more recent work do appear in the Afterword, where I reflect on current thinking about some of the issues treated in the book.

I am deeply grateful to all who made this edition thinkable, and doable, including Andra Akers, Pat Bateson, Richard Francis, Eli Gerson, Peter Godfrey-Smith, Russell Gray, Paul Griffiths, Eva Neumann-Held, Evelyn Fox Keller, Barbara Smith, Kelly Smith, Kim Sterelny, Cor van der Weele, and Rasmus Winther for their suggestions, their constructive complaints, and their intellectual and moral support.

Preface

As I look over the table of contents of this book, with its short chapters at the beginning and end and its long ones in the middle, I am reminded of the drawing of the boa in Saint-Exupéry's *Little Prince*. Having swallowed an elephant, it was slender of head and tail and grossly swollen in between. I have sometimes felt that in taking on the nature-nurture complex, I was embarking on something at least as unlikely to end gracefully as that encounter between elephant and serpent.

My interest in the various meanings of the biological can be traced back to an adolescent preoccupation with determinism. This early concern was fed by an undergraduate major in psychology, which ensured exposure to the full range of questions about nature and nurture in development, psychopathology, personality, and animal studies. Graduate studies in psycholinguistics raised the issue of innate language structures, and more recently I have watched many of these topics resurface in the wake of the sociobiology furor. The persistence of nature-nurture oppositions and the variety of forms they take have continued to fascinate me — hence, to make a tortuous path straight, this extended inquiry into our ways of thinking about constancy, ordered change, and variability.

Many readers, even sympathetic ones, will read this and look for an edifice to replace the nature-nurture complex, preferably on the same site. The demand for an alternative is legitimate, but unrealistic expectations are not, especially when they are conditioned by the very assumptions that led us astray in the past. I believe the outlines of the alternative are indeed to be found in the following pages, but I do not think they will be clearly visible to those who are searching for another grand scheme to replace the ones we have already embraced in our attempts to come to terms with life processes.

The reader who is open to a more modest approach to understanding may find much here that is congenial and familiar. I hope, though, that the person whose first reaction is "Of course" will linger long enough to see whether I have provided any novel insights into a way of thinking that may not seem new at all; much of my material is drawn from the great store of things everyone "knows." But there are many ways of knowing and of using and misusing what one knows.

I think it is instructive that the two most common reactions to my work are quite opposed in their implications. Some fear I am destroying the very basis for inquiry, and that they will be without direction if they really give up the assumptions under criticism. Others, the ones who react with "Of course," feel these issues were settled long ago. Certainly the first fear is mistaken, and this is proved by the work done by the minority of people in the second group who have not only adopted a new vocabulary but have actually implemented, in their research and interpretation, the integrated approach to development and evolution being advocated here. Though it may seem an unkind task, one of the things I try to do in this book is show that many of the people who would unhesitatingly place themselves in that last small category still cling to the very ideas they claim to have left behind.

To those who fear losing their bearings, I would say, "Take heart; there is much to be learned." To those who use an interactionist vocabulary of sorts but who have not yet fully integrated the conceptual changes required by a truly constructivist interactionist approach, my message is, "Don't stop now; the best is yet to come." And to those stalwart few who have done it all, what can I say but "Bless you"?

It has long been my perverse dream to write a book, take credit for its every virtue, and blame its flaws on the miscreants who had misadvised me. Unfortunately, this is not that book. I am aware of its shortcomings and the responsibility I bear for them; I am also aware of how much I have benefited from the criticisms and help of many people.

First I would like to thank Peter Klopfer and Pat Bateson, who broke into the autistic frenzy of my sabbatical scribbling and suggested that I think of publishing. During the past few years, they have helped me find my way through a world that was new and sometimes bewildering. They have encouraged me, read my work with care, and given me new opportunities to learn. Don Symons, my most faithful source of "Yes, buts," increased my awareness of a number of issues; that I have surely not met his challenges satisfactorily vexes me, and I am grateful that he continued our correspondence even when we seemed to write past each other more often than not. He was generous with time and praise when it helped, pointed in his criticisms when it counted. Colin Beer also offered many useful suggestions.

Conversations with a number of people helped me during the time I was thrashing out some of the issues that are discussed in this book (and some that are not); among them were Richard Burian, Margaret Clarke, Russell Gray, Howard Gruber, and Philip Silverman. These people were also valuable sources of reprints and suggestions for reading. Others were gracious enough to answer me when I asked them peculiar questions without warning, sometimes in person and sometimes by mail. I hope they are not displeased if they find traces of their thoughts in these pages, and I apologize if I have not made good use of them.

Finally I want to thank those who sustained me through these two long years of writing and thinking: friends who didn't give up on me during my extended periods of seclusion; my parents (who years ago endured life circumstances that would have undone lesser souls, and from them fashioned a warm, safe, and rich developmental system for their offspring); Mark Singleton, who was always willing to talk, to listen, to help, and to comfort, and who showed me one needn't live alone to have a room of one's own; and, of course, my children, Marina and Stefano, who have raised me with more loving permissiveness and tolerance than any parent has a right to expect.

1 Introduction

In the Western religious tradition, God created the world by bringing order to chaos. By imposing form on inchoate matter, he acted according to a convention that was very old indeed, one that separated form from matter and considered true essence to reside in the former. (Interestingly enough, "form" is cross-referenced to "universals" in the *Encyclopedia of Philosophy.*) This separation of form from matter underlies and unites all the versions of the nature-nurture antithesis that have so persistently informed our philosophical and scientific approaches to the phenomena of life. That it is not a necessary one is evidenced by studies of creation myths from other cultures (G. Bateson, 1972; Long, 1963).

Those who have argued over the origin of ideas and of biological beings have usually agreed that form in some sense preexists its appearance in minds and bodies. They have only disputed the method and time of its imposition. Most solutions to the puzzle of how form arises, therefore, including the most recent biological dogma, have incorporated the assumption that form is to be explained by pointing to a prior instance of that very form. To the extent that this is true, they are of limited value in answering questions about origins and development. Whether it is God, a vitalistic force, or the gene as Nature's agent that is the source of the design of living things and that initiates and directs the unfolding of the design thus matters little to the structure of the argument. Nor are the problems inherent in such a notion lessened by the use of a succession of metaphors, such as genetic plans, knowledge, and programs, to serve these cognitive and intentional functions.

In an increasingly technological, computerized world, information is a prime commodity, and when it is used in biological theorizing it is granted a kind of atomistic autonomy as it moves from place to place, is gathered, stored, imprinted, and translated. It has a history only insofar as it is accumulated or transferred. Information, the modern source of form, is seen to reside in molecules, cells, tissues, "the environment," often latent but causally potent. It is thought to enable these molecules, cells, and other entities to recognize, select, and instruct each other, to construct each other and themselves, to regulate, control, induce, direct, and determine events of all kinds. When something marvelous happens, whether it be the precise choreography of an "instinctive" behavior or the formation of an embryonic structure, the question is always, Where did the information come from? Was it already in the animal or the developing tissue, or did it have to be put in through learning or perhaps some embryological organizer? Was selection or instruction responsible? Is this a phylogenetic or an ontogenetic adaptation? (Was the information acquired by the species or must it have been acquired through individual experience?) The ease with which extreme nature and nurture positions are parodied ensures that no one will stand behind either straw man. No one really argues, that is, either that livers and ideas are literally in the cell or that organisms are devoid of structure, pristine pages on which anything at all may be written (and even a page has structure!). Or, to put it in negative terms, no one contends that either developmental conditions or the genes are totally irrelevant to development. Any locution that dissociates one from the straw man even minimally, however, seems to offer protection from criticism. "Encoded potential" and "biological constraints," then, allow everyone to return to work, the pesky conceptual issues behind them, peace apparently restored.

But conceptual issues are not detachable from empirical ones. They are there whenever one decides what to count as data, interprets them, draws theoretical or practical inferences from them, and frames the next research question in terms of a particular method. Meanwhile disagreement is not truly resolved, even though a generally interactionist *vocabulary* is rapidly becoming universal. The failure of this shared vocabulary to resolve old conflicts is significant; I suggest the reason is that ways of thinking about form, constancy, and change have been left largely undisturbed. (See presentations by, and commentary on, Eibl-Eibesfeldt, 1979.) At present, all that needs to be pointed out is that these exchanges often suggest, whatever the other philosophical biases of the speaker, a "preformationist" attitude toward information:[1] it exists before its utilization or expression. Some views allow assembly of information from a variety of sources, but this in turn implies that it exists in several loci before being collected;

such views thus perpetuate the mistake while seeming to correct it. In addition, information is conceived to be a special kind of cause among all the factors that may be necessary for a phenomenon, the cause that imparts order and form to matter. The alternative to such a preformationist attitude toward form is not a classical epigenetic one. Not only did this traditionally require that order arise from chaos, an unsatisfactory solution at best, but it often posited a vitalistic force as well, to effect the recurrent miracle.[2] This brings us full circle to the preexisting form, this time ready to inform the formless rather than simply waiting to unfold. Instead, it is ontogenesis, the inherently orderly but contingent coming into being, that expresses what is essential about the emergence of pattern and form without trapping us in infinite cognitive regress (where was the pattern before it got here?). A proper view of ontogeny, however, that doesn't simply resolve into one of the old ones, requires that the idea of ontogenesis apply not only to bodies and minds, but to information, plans, and all the other cognitive-causal entities (more on this phrase in the next chapter) that supposedly regulate their development. Developmental information itself, in other words, has a developmental history. It neither preexists its operations nor arises from random disorder.[3] It is neither necessary, in an ultimate sense, nor a function of pure chance, though contingency and variation are crucial to its formation and its function. Information is a difference that makes a difference (G. Bateson, 1972, p. 315), and what it "does" or what it means is thus dependent on what is already in place and what alternatives are being distinguished.

If information (along with its terminological and conceptual variants) is developmentally contingent in ways that are orderly but not preordained, and if its meaning is dependent on its actual functioning, then many of our ways of thinking about the phenomena of life must be altered. Development can no longer be explained as a combination of translated information from the genes (to make innate features) and information acquired from the environment (to modify, supplement, or complete those features). Nor can phenotypic features be divided into those that are programmed or biological and those that are not, or ranged on a continuum of relative degrees of programming, at least not if "programming" is meant to express anything more than relative predictability in a given context.

The association of biology (or, in many discussions, the genes) with the invariant, the intractable, or the inevitable is a serious error and one that must be undone. The further association of nature with biology compounds it. To say this is most emphatically *not* to deny relative invariance of some aspects: of the phenotype or developmental conservatism in any particular setting. It *is* to deny that they are the defining features of some special set of processes or characters that may be dubbed "biological" and distinguished from "flexible" ones by the role of genes in their developmental causation. It is further to deny that the common is always necessary. Nor am I disputing the possibility or utility of genetic analysis or of accounting for phenotypic differences by specific environmental or genetic variation; it is the granting of special ontological status to certain *traits* on the basis of such findings to which I object, for this leads only too easily to conclusions about nature and fate that are neither justified nor useful.

It is my contention that developmental information does "develop," not by special creation from nothingness, but always from the conditional transformation of prior structure—that is, by ontogenetic processes. Depending on the level of analysis, such transformation can be described as resulting from interactions *among* entities, such as cells or organisms, or from interactions *within* an entity, such as an embryo or a family. Since sets of interactants can be at least partially nested, a change in focus is sufficient to shift from one perspective to the next. This idea, that "information" has a developmental history, doesn't arise spontaneously from nothing, but is itself prepared by empirical and conceptual developments on many fronts: various "interactionisms" in psychology, aspects of systems theories and of Piagetian genetic epistemology, discoveries in molecular and developmental biology (though not much of the most general level of interpretation of these discoveries), cognitive psychology, and ethology. It reframes some intractable problems in these fields and offers an alternative way of conceptualizing form and causation, change and variability, normality and necessity. The notion of the ontogenesis of information in what I call *developmental systems* is implicit in much that is said and done in these sciences today, but because it usually remains implicit, it coexists all too easily with its own contradiction. It is my intention to make it fully explicit, but to do so it is necessary also to render explicit the assumptions supporting the nature-nurture complex. Once the assumptions are clear, it becomes easier to see the problems that attend them.

It is not a preference for the implicit, however, that has delayed full formulation of this concept of information, but rather a multitude of factors, including the conviction that scientific Darwinism requires the distinction between *some* notion of inherited traits or forms and *some* notion of acquired traits or forms, no matter how the distinction is made, and that the alternative is a lapse into the nether world of nonscientific Lamarckianism. Waddington remarks that "Lamarck is the only major figure in the history of biology whose name has become, to all intents and purposes, a term of abuse" (1975, p. 38). Though I do not intend to go into the question of what Lamarck really meant and what happened to those ideas in the history of biology, I think it is true that he has come to symbolize, partly by association with Lysenko and therefore with everything that is seen to be inimical to progressive science in the enlightened world, much more than he needs to. This symbolic complex has made it difficult for us to synthesize our notions of inheritance with our notions of development. My intention, then, is not to rehabilitate Lamarck or to present a Lamarckian view; scholars disagree on what such a view would be, and I will refer to him almost not at all. I do hope to show that the opposition of inherited or innate characters to acquired ones (as distinct from heritable or nonheritable variance) is based on a misapprehension of what it takes to be a properly scientific Darwinian. "If we're all Darwinians," Symons (1987) asks, "what's the fuss about?" And if we're all interactionists, again, why the fuss? I suggest the fuss arises (and sometimes doesn't arise when it should) because we haven't figured out what either requires, much less how to be both at the same time. A solution that combines encoded nature with varying doses of contingent nurture is no solution at all.

The question of precisely what is passed on in heredity, and therefore the mechanics and meaning of phylogenesis, is inevitably implicated in this conceptual reorganization. The conventional idea, that is, that organs, developmental sequences, and certain behaviors are transmitted in the genes, or that information or rules for those organs, sequences, and behaviors are so transmitted, continues to undergo progressive reformulation as successive versions are criticized and discarded. But, I submit, these changes are necessarily only cosmetic because they are not allowed to do violence to the underlying idea of inherited form or the associated belief in two kinds of development: one to manifest that form, and another for everything else. Thus we have a procession of nature-nurture oppositions based on this developmental dualism, almost as though we were generating the norm of reaction for this central concept of genetic information.

One of the problems I will address, then, is the idea of developmental information in the chromosomes. What does it explain? What is it information *about?* What important questions does it or does it not clarify? When is it invoked metaphorically, and what could it mean to treat it literally? When, if ever, does it really help us understand how the living world is constituted and what the prospects for that world are? Intimately related is the problem of the meaning of the "biological." Insofar as "biological bases" have replaced innate knowledge and structure in the vocabularies of many, they have inherited, so to speak, the implications of the earlier terms while legitimating them with the aura of methodological sanctity. But what does it mean for something to have a biological basis? What could it mean for something *not* to have a biological basis? What does this all mean for the relationship between biology as a discipline and its sister disciplines? A third problem that concerns me is interactionism itself. How does it manage to be virtually universally adopted and thus to lend itself to such radically different approaches? The suspicion is that it has become conceptually vacuous while acquiring the symbolic value of a membership badge, to be flashed upon entry into serious discussion: yes, I belong to the company of reasonable people; now let's talk about the real stuff. Though interactionism is not as innocent of substance as sometimes seems to be the case, and though, in fact, its multiplicity of meanings becomes clear when it is examined closely (see, for example, exchanges between Kitchener, 1978, 1980; and Lerner, 1980), one of the reasons it serves as an all-purpose slogan is that its users often agree on more than they realize. That set of shared assumptions, of course, about biological nature that can interact with environmental nurture is what occupies much of my attention in the pages that follow.[4]

Though a full-scale critique of interactionism will not be undertaken here, psychologists and biologists continue to explore its many faces and to debate the relative merits of each. One meaning of interaction is statistical; it refers to the interaction between independent variables in an analysis of variance. When such statistical interaction is observed, an independent variable's effects on the dependent variable cannot be accurately described without reference to another independent variable.

Though it has often led to senseless attempts to partition causal primacy according to the size of interaction terms, this kind of interaction is important to the view presented in the following chapters insofar as it directs attention to the mutual dependence of effective causes. (The importance of this principle does not, however, require the discovery of statistical interaction in any *particular* investigation. This is because results reflect the conditions chosen for that research; they may or may not produce interaction terms.) When the effects of a stimulus depend on genotype (actually, on the phenotype) or when the effects of a gene depend on the cell's location in an organism, the relativity of cause-effect relations to the context of investigation is crucial.

Another sense of interaction as I see it emphasizes the necessity of viewing transactions between an entity and its surround as aspects of a single system. When we speak of organisms altering their environments, both animate and inanimate, and being altered by their interactions, it is this mobile interchange that is being highlighted.

Hyland suggests that "interaction" has too often been treated as an answer or an explanation rather than the statement of a problem for which appropriate methods of investigation must be found. He was referring to the persons-situations debate in personality theory, which I discuss in the next chapter, but his description can also be applied, for example, to some research in behavior genetics. If analyses of variance are solutions in search of a problem, he concludes, one could also say that "the solution is the problem" (1984, p. 325; see also the commentary following Hyland's article).

In the next chapters, three fundamental phenomena of life processes will be reviewed: constancy, change, and variability. Each has been explained, completely or in part, by "nature" manifested in the genes, and each, when examined closely, reveals crucial aspects of ontogenetic processes. In Chapter 2, constancy in living things will be discussed, and so-called genetic constraints will be seen to be a function of phenotypic structure and function and of developmental context. The mutual dependence of effective causes, or reciprocal selectivity, will be described as basic to the understanding of biological forms, which are viewed as results of interactive construction, not manifestations of a preexisting plan. A consideration of change, in Chapter 3, will allow examination of the development and functional contingency of information and clarification of the way reciprocal selectivity works in ontogeny. Three kinds of variability will then be treated in Chapter 4, and the distinction between inherited and acquired characters (information, programs), on which natural selection has been said to rest, will be seen to be spurious, not only with respect to the description of individuals, but with respect to evolutionary change as well. In its place the notion of developmental systems will be advanced. This idea easily accommodates multiple developmental pathways in a way that the older idea of the inherited cannot (and constancy of pathways in a way that older concepts of the acquired cannot), allowing a consistent description of ontogeny, phylogeny, and the relations between them.

In Chapter 5, the gene as animistic metaphor, a kind of ghost in the biological machine, will be discussed in the light of the analysis presented in earlier chapters. Since the metaphor takes many forms, a number of variants will be examined, and their implications and shortcomings noted. Many of these shortcomings will be traced to confusion over the concept of control itself, which is naturally basic to ordered change.

Perhaps it is in the nature of ghosts that they should tend to linger, but this particular one, the plan or coded instructions for living beings, is an especially fascinating and tenacious one; in Chapter 6, some of its conceptual sources will be suggested and I will speculate on the reasons for its prevalence and potency. The reasons are many and varied, and I can only suggest a few; they have to do with our particular philosophical heritage and the ways we think about our own actions and experiences, as well as with the history of the biological and behavioral sciences. In a more general way the belief in genetic form and fate seems to be rooted in our desire to derive from science the answers to very old questions about what we are meant to be and what we can be, questions that cannot ultimately be resolved by the methods of biology, psychology, or anthropology. In this chapter we will look in more detail at several treatments in which the notion of the biological is central, and the conceptual work of the earlier chapters will be applied to concrete examples. Alternative ways of thinking about ontogeny as ordered change will be presented in Chapter 7, the ways in which the informative ("cognitive") and intentional or directive ("causal") functions are integrated in the material processes of the organism-environment complex. It will not do to replace the vitalistic ghost in the biological machine with a macromolecular one. The ghost, as selector of stimuli and constraint on change, is a phenotypic one; it *is* the "machine" and thus has an ontogenetic history of its own.

In Chapter 8, I will describe, in a general review organized around themes that run through the entire book, some of the implications of a view of development in which "information," regulatory systems, and controls are *generated*, in which biological constraints, rather than guiding and limiting development from the citadel of the DNA, themselves develop. This view makes possible a shift in focus from types and tokens to phenotypes in all their variety and transformational possibilities, from unilinear to multiple developmental pathways, from single- to multi-leveled reality, from predestined epigenesis to ontogenetic processes in which destiny is constructed (but never, until the moment of death, completed, and sometimes not even then), from the gene to the ecologically embedded life cycle, from the acquired-innate distinction to differentiation, interaction, and integration in developmental systems, from the search for a chimerical set of "biological bases" to a sober consideration of normality and variation, stability and change.

Finally, in a chapter on prospects for ourselves and for our science, I will argue that the nature-nurture distinction has had an impact on empirical and theoretical distinctions, and has sometimes directed research and interpretation in unprofitable ways (see Ravin, 1977, pp. 38–39, on "concepts embedded in erroneous contexts"). In addition, though, it has served an important mythological function, expressing fundamental assumptions and intuitions about our place in the world, about the sources of order and disorder, about fate, and about what it means to be human. The concept of biological nature implies a hidden truth, rooted in the past and *already there*, though it may not always be evident in the phenomenal world. This past is usually the evolutionary past of the species, but sometimes it is more individual, as when the infant is thought to reveal the real nature of the particular person to come, a nature that is discernible because it is still largely untouched by nurture. It is thus used as a representation of, and sometimes even a commentary on, what is essential about the present. But a view of possibility that is unduly restricted by projecting some version of the past into the future or that naively ignores the role of present structure in preparing, directing, or inhibiting change is incapable of doing justice to the peculiar combination of flexibility and conservatism that characterizes life processes, including our own.

My aims in this book are both critical and constructive. Most of my concerns are conceptual and metatheoretical, but this does not mean they have no empirical implications. (Readers who are drawn more to applica- tion than to principle might begin with Chapter 5, returning to the earlier exposition as the need or desire arises. The nature of the critique, however, prohibits an absolute distinction between foundation and application, and throughout the presentation I have tried to illustrate my points with useful examples.)

Wherever nature-nurture distinctions are drawn, inferences are made to all sorts of things that are thought to follow from the distinction; this is why I consider it so important to clarify both the inferences and the assumptions, along with the logic that allows them. They generally seem to escape criticism, and are even defended because they are so fully inte- grated into the methodology and theory of a number of disciplines. In fact, they have often been thought to define disciplinary boundaries. A distinction that was not thought to have great theoretical and practical im- portance, whether about the formation and fate of individuals and species or about the nature and prospects of society, could never have generated the kinds of perpetual battles, the ire and distress, that this one has. Nor could it have persisted through so many attempts to refine it, redefine it, rephrase it, or simply eliminate it.

Another problem with trying to dispense with the nature-nurture oppo- sition is that it is so widely ramified and so deeply embedded that it would be an understatement to say it would be missed. Colleagues have said, "I don't know any other way to think." That this is generally true is evi- denced by the way people declare their rejection of the nature-nurture opposition, sound the rallying cry of "interactionism," then proceed to discuss genetically encoded instructions and environmentally shaped traits. Though I am acutely aware of the weary impatience evoked by nature-nurture discussions, I will refrain from offering the apologies that typically precede them. I am convinced there is another way to think, and that this other way, though it requires reworking not only our ideas about genes and environment but quite a bit besides, gives us both more and less than the old way. It gives more clarity, more coherence, more consis- tency, and a different way to interpret data; in addition it offers the means for synthesizing the concepts and methods of evolutionary biologists and developmentalists, groups that have been working at cross-purposes, or at least talking past each other, for many decades. It gives less, however, in the way of metaphysical guidance on fundamental truth, fewer conclusions about what is inherently desirable, healthy, natural, or inevitable, and this accounts for a good deal of the resistance it has met.

2 The Origin and Transmission of Form:

The Gene as the Vehicle of Constancy

Powerful and protean, and far from being banished from secular science, the argument from design is ubiquitous. Complex design, it claims, is evidence of creative, typically divine, intelligence. Perhaps because we are creatures whose existence and survival depend on our ability to discern regularities in our surroundings and in turn to leave our mark—our design—on them, we tend to infer prior design or intent from observed regularity. We formulate, that is, a descriptive rule, which is a form of knowledge, and infer from it a prescriptive rule separate from the processes we see and controlling them. It is a short step from the statement that we are as we are, and think as we do, *because* of our essential nature as biological creatures to a variant of the argument from design: our nature is created by a genetic plan, an intelligence in the chromosomes, which was in turn created by natural selection. By slow evolutionary winnowing, Nature placed knowledge of herself in the chromosomes of her creatures; this knowledge is instantiated in anatomical structure, physiology, and instinct, and—for the more refined—biological boundaries, genetic predispositions, facultative responses, and the like. Just as we place a man in the head to receive and interpret sensation and to issue commands to limbs (Ryle's ghost in the machine [1949]), so do we place a plan in the man that assembles and controls *him*. Jacob observes, "The organism thus becomes the realization of a programme prescribed by its heredity. The intention of a psyche has been replaced by the translation of a message" (1973, p. 2).[1] In our very vocabulary, form and intent, pattern and control are fused. Consider "design" as pattern and as intention, or "pattern" as noun and verb. It is just this step, short and habitual as it is, between descriptive and prescriptive rule, between observed regularity and prior plan as guarantor of that regularity, that I think we should resist. Because this metaphor of a prior plan seems not only harmless but useful and even necessary (one could argue, as Kant did, that this is the only way we can approach living phenomena), it seems difficult to object to it, but I believe it is often misleading. Because we are unaccustomed to thinking in other ways, it invites literal interpretation, and thus the illusion of having given a causal account (and even Kant argued that the attribution of intentionality to natural processes did not constitute explanation [1949, p. 309]). Once taken literally, the notion of preexisting prescriptive rule commits us to a view of development that is at once incoherent and pernicious. It is incoherent because it is inconsistent with the phenomena it seeks to explain. It is pernicious because it is just this conception of developmental causation (that some development proceeds by genetic rules and some does not) that undergirds the opposition of biological to cultural processes, the mare's nest of biological determinism and the whole nature-nurture complex.

A. Homunculoid Genes

Not surprisingly, the argument from design is no more satisfactory at the microscopic level than at the cosmic. We shall see that when processes at the gene level are examined, there arises once more the felt need for a directing force, especially since genes are only physiologically functional at certain times and in certain tissues. This time, however, the analyst is trapped. Unable to postulate another entity inside the gene or the genome (though invocation of regulatory or switch genes to explain such dynamic patterning suggests that this strategy is an attractive one), he or she may resort to more and more elaborate metaphors, which, while expressing a genuine and important intuition about ordered regularity (its air of integration), can be helpful only if they synthesize what I call the "cognitive" and "causal" functions with the material one. This must take place in the living organism and its ongoing processes, not in an ever more complicated set of metaphorical plans and images to assist the creator homunculus in its labors.

The cognitive function corresponds roughly to Aristotle's formal and final causes. The linking of formal (pattern) and final (goal or end) causes is accomplished in modern biology by the gene as source of organic form and as product of evolution. More generally, the designer usually imposes on his or her creations forms that fit his or her ends, and though few students of nature believe evolution itself is purposive, it is still seen as instilling purposes in its products. When development in turn is seen as genetically caused, the genes also serve as Aristotle's efficient causes, the ones most nearly resembling commonsense ideas of causes as effectors of change—hence "cognitive-causal" functions of the gene. What remains in Aristotle's fourfold system is the material cause, the matter to which form is imparted; this seems to correspond fairly well to the "raw materials" supposedly organized by the genes during development. Toulmin and Goodfield (1962, chapter 4) argue that the adoption in later centuries of the word "cause" for all these explanations was unfortunate because only the efficient cause corresponds to the modern sense. The "four becauses," they suggest, would have been a more accurate rendering, for Aristotle was concerned with different ways the question "Why?" can be addressed.

Whether one traces our present conceptions of life to Aristotelian notions of form and matter (see Hailman's 1982 discussion of ideas of biological causality) or, as Toulmin (1967) does, to Newtonian and Cartesian definitions of matter as inert and therefore as requiring outside animation, it is clear that our preoccupation with organisms as material objects whose design and functioning must be imparted to them has a long and complex past (see also Lewontin, 1983; Montalenti, 1974). Toulmin sees in seventeenth-century physics the mechanistic assumptions that necessitated an animistic physiology and a dualistic psychology. This is because the motivating mind or soul cannot be part of the machine it moves, which is, by definition, incapable of movement. It certainly seems that our contemporary tendency to think of causes in terms of *forces* makes it virtually necessary to collapse Aristotle's first three factors into a guiding, constructive force.

The discovery of DNA and its confirmation of a gene theory that had long been in search of its material agent offered an enormously attractive apparent solution to the puzzle of the origin and perpetuation of living form. A material object housed in every part of the organism, the gene seemed to bridge the gap between inert matter and design; in fact, genetic *information,* by virtue of the meanings of *in-formation* as "shaping" and as "animating," promised to supply just the cognitive and causal functions needed to make a heap of chemicals into a being.

Placing the design into the animal, unfortunately, does not explain the fulfillment of the design, as many have noted (Hinde, 1968; Klopfer, 1969; Lehrman, 1970; Oppenheim, 1982), and indeed, trying to explain the development of form by invoking blueprints and programs in the nucleus, as we shall see, raises more problems than it solves. That these problems are variations on themes that have been playing for centuries shouldn't surprise us; our fundamental assumptions about form and matter, stability and change, have not been much altered by advances in molecular biology. Wherever the cognitive-causal function is housed, that is, whether in the mind of God or in Drieschian vital energies, or in little computer programs in the cell, as long as it is seen as preexisting and separate from its material realization, it cannot illuminate life processes. Resolution is therefore not to be found by opposing an "environmental" or "empiricist" formulation to a "nativist" one, for, as noted above, this merely shoves the cognitive-causal function outside again, without altering the assumption that order is explicable by the imposition of form on the formless and of movement on the inert.

The logical similarity of these traditionally opposed views is strikingly highlighted when, in attempting to convey the interdependence of biological processes even at the intracellular level, where "cytoplasm has its own developmentally significant properties," Grobstein remarks, "although important determinants for development lie in the zygote nucleus, the cytoplasm of the zygote is not simply a blank page on which the nucleus prints out instructions" (1974, pp. 110–111). That the tabula rasa of the empiricists is so easily translatable into naive nativism's conception of an "outside" completely organized by the "inside" is perhaps not too surprising; perhaps it is not disturbing, either, since the canon of enlightened "interactionism" has long decreed that both extremes are nonsense, that right thinking embraces nature *and* nurture. What is not so clear is the extent to which the kinds of arguments that make the extremes untenable persist in these conventional interactionist arguments. This is true whether interest is focused on the development of form in ontogeny or in phylogeny. In fact, the habit of thinking about phylogeny and ontogeny as *alternative processes whereby information enters the organism* is the very frame on which our endless nature-nurture disputations are woven. Nativism and empiricism require each other as do warp and weft. What they share is the belief that information can preexist the processes that give rise to it. Yet information "in the genes" or "in the environment" is not bio-logically relevant until it participates in phenotypic processes. *It becomes meaningful in the organism only as it is constituted as "information" by its developmental system.* The result is not *more* information but *significant* information.

Few question the background assumptions that biological form can originate in two ways, by being inherited or by being acquired, that the genes, as vehicles of inherited form, are the guarantors of phylogenetic constancy (and eventually, change) through the manifestation of these forms by special developmental processes designated "maturational" (or more recently, by epigenetic rules and canalization), that the course of individual lives is some sort of amalgam of what arises spontaneously from within and what is sculpted from without. This is the case even when particular controversies (how much of the nervous system is prewired, what part of social behavior must be learned, etc.) may be spirited and noisy.

In addition to emphasizing the structural similarity of empiricist and nativist conceptions of organic formation, Grobstein's comment on the cytoplasm as blank surface awaiting the genetic inscription should make clear the way in which these conceptions have been progressively shunted down the analytic levels to intracellular processes. Hailed as the long-awaited synthesis of all the venerable antitheses in the study of life (Jacob, 1973, p. 2), legitimized by breathtaking advances in molecular biology and by integration with a well-elaborated vocabulary of information systems technology, the genetic code as keeper and executor of phylogenetic in-structions seems to occupy a position too central to the life and social sci-ences to question. Yet, if pressed, any student of life processes will admit that the species-typical genotype guarantees the species-typical pheno-type only if the environment maintains its typical parameters as well. This is so obvious as to merit nothing but an impatient sniff: unquestionably the genes require the proper supporting conditions to do their work, but when those conditions are present, the genotype supplies the fundamental pattern of the organism.

Sometimes it is useful to question the unquestionable. How does one know, we might ask, that it is the gene that supplies the species pattern? Answer: because when the genotype changes, within or between species, so does the phenotype, and sometimes we can actually trace the sequence of physiological changes by which this occurs. The same analytic justification evidently does not, however, allow us to conclude that species pattern is contained in the environment when we observe that environmental differences frequently bring about phenotypic ones, even though causal sequences can sometimes be traced in these cases as well. In fact, pattern is often maintained despite considerable changes in genotype, environment, or both. It is possible that in many organisms a great deal of genetic material has no phenotypic expression at all (Doolittle, 1982). Whatever the outcome of the debates over "junk" and "selfish" DNA, there is ample evidence that many gene duplications and alterations are phenotypically silent. When environmental variation leads to no phenotypic variation, the phenotype is typically concluded to be "under genetic control." In contrast, when genetic variation fails to lead to phenotypic variation, people don't usually say that the species-typical phenotype is under *environmental* control. The conviction remains that there exists a hierarchy of causes, some quite lowly, involving only the crudest of constraints (not causes at all, really, but raw material), and others that are the true source of form.

It is easy to demonstrate that a kind of causal symmetry exists whereby genetic and nongenetic factors alike can be sources of variation in form, and whereby constancy of form generally requires relative constancy in developmentally relevant aspects of both genome and environment. The constant, or typical, form is then attributable to both (Oyama, 1981, 1982; Waddington, 1957, p. 186). Yet this symmetry is generally ignored. Abnormality can certainly have either genetic or nongenetic origins, people readily agree, while *normality* is believed to be a result of the homunculoid gene's ability to organize matter into living beings. As Lewontin (1983, p. 35) somewhat wryly remarks, one could see the "penetration of the bourgeois world view into biology" in the distinction between intellectual labor (gene as design) and "mere production" (enzymes, cell structure, etc.).[2]

While it is readily acknowledged that the form a biological raw material takes depends on the genotype in question, intellectual fairness requires that we also admit that the form a gene influences, if any, depends on the available constituents and the developmental context, including the rest of the genome. More generally, what a cause causes, whether it be material cause, formal cause, or some other variety is contingent, and is thus itself "caused." What I am arguing for here is a view of causality that gives formative weight to all operative influences, since none is alone sufficient for the phenomenon or for any of its properties, and since variation in any or many of them may or may not bring about variation in the result, depending on the configuration of the whole. (See Taylor, 1967.) In what will be referred to here as the reciprocal selectivity of influences, or the mutual dependence of causes, not only does an entire ensemble of influences contribute to any given phenomenon, but the effect of any interactant depends both on its own qualities and on those of others, often in complex combinations. This is consistent with Lewontin's description of constrained relationalism (1983), as well as with much of systems thinking (see Waddington, 1960, pp. 79–83, on organized causal networks; G. Bateson, 1972; von Bertalanffy, 1967, 1972), and is much more faithful to the observations of molecular biology, embryology, and the behavioral sciences than are the nature-nurture models or most versions of "interactionism," a brave new view (or rather, conglomeration of views still struggling for coherence) that in most cases promises more conceptual improvement than it delivers.

B. Persons and Situations

The debate over whether human behavior is primarily a function of personality (internal factors) or of situations (external factors), for instance, is an old one in psychology. In a revision of her much-reprinted feminist analysis, "Kinder, Küche, Kirche as Scientific Law: Psychology Constructs the Female," Naomi Weisstein (1971) declares that psychology knows nothing about women's real characteristics, needs, and wants, despite much theorizing about feminine personality, and that this weakness extends to theories of personality in general. This is because such theories have looked for inner traits rather than at the power of social expectations and situational demands. Her declaration that psychology is useless for understanding women is ironic in light of the fact that her critical points are all made with the findings of psychological research, but the point here is that the inner-outer causal dichotomy is joined with the issue of interindividual variation and constancy. This is not just Weisstein's construction; it also reflects the way the fields of personality and social psychology have tended to define themselves. What social psychological research shows, however, is not creatures pushed and pulled by external forces as opposed to internal ones, but the workings of largely shared perceptions, motives, and beliefs in particular circumstances.

One of the consequences of conflating inner and outer causes with individual differences and similarities is that people are often seen to be the source of their own behavior only when they are not behaving like others. (Another is that it is inconsistent with the habit, often found at an another level of analysis, of conflating inner causes with *similarities* among individuals, while many individual differences are attributed to outer causes. This is, of course, the level at which the biological nature of a species is defined.) I have always wondered why, for instance, if it can be shown that most or all people would have done what X did if they had been in X's position, X is not the responsible cause of his or her actions, the situation is. This way of looking at life may be congenial to our individualistic ethic but it has some paradoxical implications. If inner causes explain little about human behavior, and outer ones explain all, what does this mean for just that individualistic, self-determining ethic (and doesn't that ethic risk becoming just another external cause)?

The idea of responsibility is very much involved in this muddle. It is surely not accidental that on the first page of the collection of feminist writings in which Weisstein's paper appears, the epigraph begins, "The fault lies not in our stars, our hormones, our menstrual cycles or our empty inner spaces, but in our institutions and our education." If one wants to divest oneself of blame for a devalued characteristic, one can blame something(one) else. But, and this is especially poignant if one of the devalued characteristics is passivity, how then does one reconstruct oneself and one's world? One study showed that rape victims who accepted some responsibility for the incident were less depressed than those who accepted none, apparently because they felt less helpless, and thus less passive and vulnerable (Brody, 1984).

Surely the way to encourage people to think about and improve their lives is not to replace one set of coercive determinants with another; and surely the way to think about responsible action is not to juggle inner and outer, ultimate and proximate causes, and hope that reasons and responsibility will miraculously squeeze through some narrow space where all these causes collide in persons.[3]

In a 1973 "interactionist" critique of the persons-situations issue in personality theory, Kenneth Bowers declares that the two are "codeterminers of behavior, and they need to be specified simultaneously if predictive accuracy is desired" (p. 322), and that "situations are as much a function of the person as the person's behavior is a function of the situation" (p. 327). Such a view is clearly intended to rectify the weaknesses of the traditional approach. He seems, however, to retain crucial aspects of the dichotomy he is attacking. He associates trait theories with prediction of behavioral stability across situations and individual differences within situations, whereas situationism predicts behavioral differences across situations and treats individual differences as error. This is historically accurate. But to the extent to which persons and situations are treated as alternative explanations for behavior, as opposed to possible sources of variation, the interactionist principles Bowers offers are violated. At one point he argues that failure to change a behavior should be seen as evidence for the importance of intrapsychic factors in perpetuating behavior. Since there is no behavioral variance to partition here, he seems to be arguing for some other kind of cause. He further asserts that the amount of effort required to change a behavior indicates the "degree to which the behavior in question was originally attributable to characteristics of the person instead of to evoking and maintaining conditions" (p. 325). This inference is difficult to defend. The effectiveness of a kind of variable in *changing* a phenomenon does not necessarily tell us anything about whether that kind of variable was important to the *development* of the phenomenon. If it did, we would have to conclude that because cement does not dissolve in water, water was not important in making cement, or that because feeding an adult plant may not increase its height, nutrition or other "environmental factors" were not important to the growth of plants or to the determination and maintenance of their adult height. Worse, attributing *behavior* to persons or to conditions makes as little sense as does attributing any phenotypic characteristic to varying proportions of inner and outer causes.

What Bowers is confronting is the same conceptual confusion we encounter in the nature-nurture controversy, not the least of which involves the identification of stability with the inside and variability with the outside. The inadequacy of the internal-external distinction becomes especially evident, as we shall see, when "internal factors" seem to generate change rather than inhibiting it, as they do in the persons-situations debate.

There is more than structural similarity here; Bowers says that the explanation of individual differences leads directly to a "biological or genetic point of view" that postulates the existence of organismic differences before situational differences can have any impact (p. 327). Though this is indeed the way the argument tends to go, and though it is this chain of inferences that Weisstein was both accepting and fighting, some comments are in order. First, Bowers seems to be claiming some sort of developmental primacy, in the senses of both priority and importance, for "genetic" traits, which again seems at odds with an interactionist synthesis. It is, furthermore, neither valid nor necessary to his argument. Identical or similar genomes may have different environments, intra- or extracellular biological situations, if you will, and this uniqueness of milieu may be present as early as uniqueness of genotype. Even if it could be shown that early environments were identical and diverged only later, it is not clear what theoretical significance this would have. Uniqueness is indeed an important fact of life, and would be whether all or no differences were attributable to variation in early environment. (The relation of evolution to heritable variation is another matter.) It is just the importance of uniqueness that many trait theorists have tried to emphasize. Basing the argument on an opposition of biology to environment helps only insofar as a decidedly noninteractionist view of psychological and developmental processes is adopted.

Phenotypic stability, like phenotypic change, must be explained by examining organism-situation relations, not by attributing it to one or the other factor. Bowers may have been scoring a rhetorical point against situationism, but if so, it was not a point in favor of a thoroughgoing interactionism.

Just as Weisstein, in battling one kind of determinism, substituted another and thus burdened her argument with some of the very assumptions she might well have questioned, so Bowers adopts a notion of genetic causation that weakens his interactionist thesis. No one can do everything, and my comments in these cases, as in many others in this book, are labors of love, however critical I may seem. I am very much in sympathy with the intellectual—and, in Weisstein's case, political—thrust of these treatments, but just because of this sympathy, am dismayed at the inconsistencies that weaken them. One of the first orders of business, and by no means the easiest, is the separation of theoretical points from their political implications (if any).

The subsequent course of the persons-situations debate will not be followed here. As is the case with a number of my examples, these were chosen more for their usefulness in illustrating certain points than for their recency. That these issues have not been resolved in psychology is confirmed by a glance at any recent text or journal, where traits continue to be ascribed, after an obligatory "interactionist" introduction, to mainly biological or mostly cultural factors; to genetic predispositions and physiological bases or to experience; to phylogenetic or individual history; just as people's behavior continues to be attributed to internal or external factors. It is instructive, though, to notice the relations of the persons-situations opposition with the problems we have been discussing.[4] Other dichotomies, including the venerable Great Man–historical determinism distinction in history, as well as the related one between history as a humanity and as a social science, or between cultural and geographical determinism in anthropology, could be analyzed in similar ways. What strikes one is the relative slowness and internal confusion with which an interactionist perspective emerges, and its ability to coexist with its contradictions. (For an update on interactionism or "contextualism" in psychology, see Lerner, Hultsch, and Dixon, 1983.) In addition to the kinds of internal inconsistencies noted in Bowers's presentation, which, I might add, was very astute and had much to commend it, there are incongruities among levels of analysis as well. A Meadian social psychologist, a Piagetian, or an ecological anthropologist may never question the inherited-acquired distinction, for instance, or may question it for behavior but not morphology, or for human behavior but not that of other animals.

An entire research tradition in social psychology treats humans as scientists analyzing events and attributing these to causes of various types (Heider, 1958). One aspect of this process involves attributing behavior to variable or stable factors that are either internal or external to actors. Such judgments may be made about one's own behavior as well as that of others; self concept, then, can be a partial function of these causal analyses. It might be interesting to apply these ideas from attribution theory to some of the controversies we consider here.

In a developmental analysis of male-female personalities and relationships in humans, Dorothy Dinnerstein (1977) proposes that gender differences, which have commonly been conceptualized either as biologically necessary or as imposed by society at large, are a result of the child's experience of being nurtured by a female. Such an approach, since it examines a virtually universal aspect of human life, is perfectly consistent both with the possible cross-cultural generality of certain features of gender and with an exquisitely finely tuned and complex view of human perceptions, emotions, and interactions with the social world. Though it is frequently assumed that what is learned is relatively easy to alter, neither Dinnerstein nor any thoughtful reader would claim that this fact of our lives, or its consequences, would be simple to change.

C. Intrinsic Meanings?

Since the genome represents only a part of the entire developmental ensemble, it cannot by itself contain or cause the form that results. But then, neither can its surroundings. As is frequently the case in these matters, people in some sense know this perfectly well, and say so. They often end up betraying their own good sense because they view a lack of variation (within the organism if focus is on individual nature and within the species if focus is on species nature) as evidence of inherent, necessary qualities. Hofstadter points out, in a discussion of messages and their decipherment, that if a deciphering mechanism is universal, then it is easy to attribute meaning to the message itself. "Meaning is intrinsic if intelligence is natural" (1980, p. 171; notice the use of "natural" for "universal"). If the same message can be interpreted in several ways, it is harder to consider the several meanings as inherent in the message, though I suggest that this is precisely what behavioral scientists are attempting to do with concepts of genetically encoded potentials. Weight, Hofstadter points out, was ascribed to objects until we realized that it varies with gravitational pull; now we conceive of mass as invariant, and ascribe *it* to objects instead of weight. All messages are coded, he asserts, but when a code is very familiar one tends to ignore the decoding mechanism and to identify the message with its meaning (p. 267). This is just what seems to have happened with the genetic code, its phenotypic "meanings," and the "decoding mechanism" of ontogeny, and this is why the variety of onto-genetic pathways, and therefore of decoded meanings, is so important. In Chapter 4, we will discuss such variety, which refocuses attention on the processes that, by "pulling" different information out of the genome in different order, or, better, by generating different information, depending on developmental context, create the variety of phenotypic outcomes we observe or might observe, conditions permitting. By so doing, they make it more difficult to identify a single meaning with the genes, how-ever common that outcome might be or however enduring it might seem in individuals. Because we often think in terms of single forms rather than sequences of related states, in fact (a tibia is just a tibia, we say, even though an expert can distinguish innumerable developmentally distinct shapes and sizes, not to mention individual differences; and an introvert is an introvert, even though we actually define introversion by abstract-ing from many actions in many situations), we are tripped up by our own precious generalizing faculties and ignore differences that are basic to the phenomena we are describing.

While clear-cut alternative phenotypes or experimentally produced variation make Hofstadter's point about the relationship between inher-ency and reliability in a striking manner, the everyday regularity of matu-ration does it in a subtler way, confronting us with the differentiation of multiple "meanings" from the structure and function of a single eco-logically embedded cell. In any instance of gene transcription, the gene "selects" from the available chemical constituents and is "selected" or activated by other factors. These other factors may include those very con-stituents, or gene products. This logic holds for regulatory elements that may exert important influences without being transcribed; their effects depend on their position vis-à-vis other genes and on nongenomic fac-tors as well (Raff and Kaufman, 1983). Causation is multiply contingent, and influences both select each other and determine each other's effects. When biologists describe particular chemical, tissue, or organism interac-tions, this reciprocal selectivity is evident. Yet constancy, in the individual over time or across individuals, goes on being attributed to the genetic program, which functions as a mindlike entity controlling organic processes and containing the plan of the complete organism. (It is instructive to compare the attributes of the genome in current theories to those of the life force in vitalism. See, for example, Beckner, 1967.) While it may do no particular harm, furthermore, to use "genetically programmed" as shorthand for "reliably present in the species," we often invoke such programming to make predictions in situations where we know that the typical genome, but not all the typical conditions, will be present. Unless we know that the new conditions include everything needed for the development of the feature in question, this seems fairly foolhardy.

A view of single causes bringing about "their" effects encourages us to make this kind of projection from partial knowledge, whereas a more inclusive view that embraces configurations of causally relevant factors, some more stable than others, makes it difficult to ignore less salient, but still causally important, aspects of a complex system. This latter view allows us to avoid some of the logical problems raised by unitary conceptions of causality. In doing so, we do not give up our ability to account for the impressive regularity of many developmental processes; the initial developmental system is a prime producer of its own subsequent states, and the more uniform these early systems are, in general, the more uniform the ontogenetic sequences will be. The richness introduced by the notion of an interactive system, as opposed to a static causal entity, provides for *both* the predictability of much ontogenesis *and* the intra- and interorganismic variability that is so basic to the living world (more of this in Chapter 7).

Yet "biological" characters are customarily distinguished from "phenotypic modifications" (modifications of *what,* one might ask) as the essential is distinguished from the accidental. It is our essentialist viewpoint, I think, that allows us so easily to confuse *change* in an existing feature with *variation,* a confusion that allows us to say that a biological trait is not (easily) modifiable and to mean both that it can't (easily) be changed in individuals and that it is (nearly) uniform among individuals. Such usage reflects and perpetuates the belief that variation is *deviation* from an internal ideal.

In actual phenotypic processes, whether at the biochemical or the behavioral level, all these influences are integrated and their biological meaning is determined. When their presence and mode of interaction are constant or equivalent, so is the phenotypic result; when they vary, so may the phenotype, though such processes are frequently redundant and conservative, preserving constancy of outcome despite considerable fluctuation in interactants.

Form emerges in successive interactions. Far from being imposed on matter by some agent, it is a function of the reactivity of matter at many hierarchical levels, and of the responsiveness of those interactions to each other. Because mutual selectivity, reactivity, and constraint take place only in actual processes, it is these that orchestrate the activity of different portions of the DNA, that make genetic and environmental influences interdependent as genes and gene products are environment to each other, as extraorganismic environment is made internal by psychological or biochemical assimilation, as internal state is externalized through products and behavior that select and organize the surrounding world. If biological plans, constraints, and controls have a serious meaning, it is only in such mobile, contingent phenotypic processes, not in a preformed macromolecular code specifying the species type, of which type the individual is but a token.

Organismic form, then, constant or variable, is not transmitted in genes any more than it is contained in the environment, and it cannot be partitioned by degrees of coding or by amounts of information. It is constructed in developmental processes. What are transmitted are *macromolecular form,* which, though it is necessary for the development of phenotypic form, neither contains it nor constitutes plans for it, *and developmentally relevant aspects of the world.* Chromosomal form is an interactant in the choreography of ontogeny; the "information" it imparts or the form it influences in the emerging organism depends on what dance is being performed when, where, and with whom. The dance continues throughout the life cycle, and everything that occurs in that cycle, from the first moments to the moment of death, from the most permanent structure to the most evanescent, from the most typical feature to the most divergent, is constructed from these interactants. In the individual, constancy of form is largely ensured by the fidelity of chromosome duplication and by the fact that the maintaining influences for any given process or structure are apt to be supplied by the organism itself or by its niche, which includes other organisms. Cells in vitro may dedifferentiate, but in real life they generally remain safely embedded in the vital network that sustains them. When that network is disrupted through injury, cells may dedifferentiate in vivo as well, allowing regeneration of parts. Many other interactants, of course, are sought, found, or created by the organism itself in due time. Over evolutionary time, these processes have become more or less reliably coupled; constancy across generations and within a population is due to these relations, whose reliability is variable. There is no vehicle of constancy (even though the coined term, "interactant" may have an unfortunate particulate connotation), unless the organism and its niche, as they move along time's arrow, are so conceived. The developmental system, however, does not have a final form, encoded before its starting point and realized at maturity. It has, if one focuses finely enough, as many forms as time has segments.

Constancy is constructed. To understand it, then, we must look at its processes, not only at its constituents, and be ready to meet them on as many levels as their complexity justifies.

3 The Problem of Change

In thinking about forms, we tend to focus on constancy rather than change
—constancy of a character through time, in spite of turnover in constituent materials and shifting conditions, similarities among related individuals or members of a species, continuity across generations or among
related species. In fact, though, change and variability are as basic to biological processes as uniformity. Change and variability are not the same
thing, though they are associated in important ways, and they both pose
difficult problems for the conception of the directive gene. We will consider change in an entity first; variability is the subject of the next chapter.

Embryogeny, as orderly, recurrent sequences of change, is the perfect
synthesis of constancy and change. From day to day, sometimes from
minute to minute, the entire system, increasingly intricate, changes—
forms shift and disappear, lift and fold, divide and spread—yet each phase
is largely predictable, at least in its outlines, as its end.[1]

A. Preformation, Epigenesis, and Golden Means

Students of development in the seventeenth, eighteenth, and nineteenth
centuries discovered more astonishing complexity of change than had
been previously imagined, and it was then that the preformationist and
epigenetic views became engaged in an increasingly subtle dialectic, co-
alescing by the late nineteenth century to form what Oppenheim (1982,
p. 37) describes as an essentially modern position: "transmission of a *predetermined* (or rather, a preorganized) germ plasm in the nucleus that in
the course of ontogeny is expressed via cytoplasmic *epigenesis.*" It will be
recalled that Oppenheim uses "epigenesis" for progressive ontogenetic
change. It is also important to bear in mind that the course of development
was agreed by preformationists and epigeneticists alike to be predeter-

mined in some sense; the question was how this was to be understood. Gould says:

> The solution to great arguments is usually close to the golden mean, and this debate is no exception. Modern genetics is about as midway as it could be between the extreme formulations of the eighteenth century. The preformationists were right in asserting that some preexistence is the only refuge from mysticism. But they were mistaken in postulating preformed structure, for we have discovered coded instructions. (It is scarcely surprising that a world knowing nothing of the player piano — not to mention the computer program — should have neglected the storage of coded instructions.) The epigeneticists, on the other hand, were correct in insisting that the visual appearance of development is no mere illusion. (1977, p. 18)

I, on the other hand, tend to be skeptical of golden mean solutions, because when two great traditions battle for long periods, there is generally, along with their disagreements, some basic misapprehension that they share, and from which, therefore, they do not attempt to dissuade each other. Preformationism did offer a refuge from mysticism in the sense of vitalistic forces acting directly on development each time it occurred, but only at the price of postulating a one-time-only simultaneous creation of all things (and thus no need for real development). Though contemporary thought rejects nested bodies, it does not often balk at preformed targets and instructions. The preexistence that the preformationists, the epigeneticists, and current orthodoxy all agree upon is form, whether miniaturized and encapsulated, re-created by a vitalistic force, or inscribed on a molecule. The corollary is that the preexisting form must be the "same" as the final one; for the preformationists both were concretely material, for the epigeneticists form was first disembodied, then embodied, and for most modern thinkers it is initially material but cryptic, then manifested in the phenotype. In fact, preformationism was not as naive as is often claimed; the important versions during biology's early years acknowledged differences in the relative position and shape of encapsulated organs, agreeing that if the embryo were enlarged it would not look like a finished organism (Gould, 1977, p. 20; Oppenheim, 1982). This brings them closer both to their traditional opponents and to modern views.[2]

Though Gould states that the preformationists were in error in postulating preformed structure, there is really nothing problematic about this idea. What is misguided, not only about the preformationist belief but the modern version as well, is the assumption of *correspondence* between initial and final structures. The chromosomes are indeed highly structured, as are the cell organelles, the chemical substrates, and the extracellular environment.

Emphasis on the structure in the genome without full acknowledgment of structure in the surround is common. Gould (1977, pp. 21–22) claims that the preformationist critique of epigenesis is still valid: if the egg is "truly unorganized, how could it yield such consistent complexity without a directing entelechy. It does so, and can only do so, because the information—not merely the raw material—needed to build this complexity already resides in the egg." Preformationists and epigeneticists agreed that a formless egg required form from without. Preformationists placed accomplished form inside the egg, while epigeneticists rejected this solution and instead posited an additional force. Gould appears to accept the vision of the formless egg and to place the entelechy-as-program inside. (See also Mayr, 1982, pp. 105–106.)

The situation is thus quite peculiar. What scientists say in some contexts is contradicted by what they say, know, and do in others. It is therefore difficult to grasp what is happening without following arguments or trains of thought quite closely, and this is one of my tasks in this book. In fact, no biologist seriously limits structure to the chromosomes; they sometimes sound as if they do because *they assign formative relevance only to the DNA, where the encoded representation of the phenotype (or of the instructions for building it) is thought to reside.* This is the error—along with the associated idea that unless such a representation exists, development cannot be structured—and it is a pervasive and fundamental error indeed.[3]

If matter and form are distinct, as one part of our classical tradition tells us, and matter is inert (recall Toulmin's argument about seventeenth-century thought [1967]), then both the emergence of form and change in that form must be explained. We tend to see the "biological" as that which has the power to effect change without being changed. Maturation, conceived causally rather than descriptively, is seen as a force bringing basic characters into being, without requiring, indeed, without permitting, more than minimal environmental influence (Oyama, 1982). Maturation in turn is seen as driven and guided by the genes, which "initiate and direct" development (Gesell, 1945, p. 19), which are potentially immortal (Williams, 1974, p. 24), and which, of course, are not generally altered by their cellular interactions. To continue the commonly agreed upon causal progression, the genes are formed by random variation and natural selection. Gatlin notes that "the words 'natural selection' play a role in the vocabulary of the evolutionary biologist similar to the word 'God' in ordinary language" (1972, p. 164), and Montalenti observes that natural selection is the biological *primum movens* (1974, p. 13). Just as traditional thought placed biological forms in the mind of God, so modern thought finds many ways of endowing the genes with ultimate formative power, a power bestowed by Nature over countless millennia.

"But wait," the exasperated reader cries, "everyone nowadays knows that development is a matter of interaction. You're beating a dead horse."

I reply, "I would like nothing better than to stop beating him,[4] but every time I think I am free of him he kicks me and does rude things to the intellectual and political environment. He seems to be a phantom horse with a thousand incarnations, and he gets more and more subtle each time around. Just look at the horselets, infrahorses, and metahorses described in this book. What we need here, to switch metaphors in midstream, is the stake-in-the-heart move, and the heart is the notion that some influences are more equal than others, that form—or its modern agent, information—exists before the interactions in which it appears and must be transmitted to the organism either through the genes or by the environment. This supports and requires just the conceptions of dual developmental processes that make up the nature-nurture complex. Compromises don't help because they don't alter this basic assumption."

Jacques Monod, who, with Jacob, presided over some of the most exciting developments in early molecular biology, describes in minute detail various macromolecular processes and their organizing functions. When he engages in straightforward description, the complexity and interdependence of causes are clear. When he interprets these processes in more general terms, however, an interesting thing occurs. He says, for example, that the genome "entirely defines" protein function, and asks if this is contradicted by the statement that the protein's three-dimensional structure has a "data content" that is "*richer* than the direct contribution made to the structure by the genome." His answer is that, because the three-dimensional, globular structure appears only under "strictly defined initial conditions," only one of all possible structures is realized. "Initial conditions hence enter among the items of information finally enclosed within the globular structure. Without specifying it, they contribute to the realization of a unique shape by eliminating all alternative structures, in this way proposing—or, rather, imposing—an unequivocal interpretation of a potentially equivocal message" (1971, p. 94). But if initial conditions select one folded structure among an array of possible ones, thus contributing to the unique shape, they *do* specify it, in cooperation with the linear structure. The particular globular shape results only when particular chains fold under particular conditions. Monod is forced, in his terms, to admit that the structure therefore "contains" more information than it would if such conditions were not critical. But his commitment to the power of the gene entirely to define leads him to withhold "specifying" power from the cellular environment. This is not due to idiosyncratic use of "specify"; elsewhere he makes much of the *specificity* of enzyme action—that is, the selective interaction with only one or two molecule types from many.

Earlier in the same book (p. 84), he asserts that he uses "epigenesis" for all structural and functional development, not in the old sense of the epigeneticists, who, in contrast to the preformationists, "believed in an *actual* enrichment of the initial genetic information." Since the eighteenth and nineteenth centuries lacked the contemporary notion of genetic information, one can only speculate on the precise meaning of this assertion, but in the light of Monod's general position it seems reasonable that he was distancing himself from any view that attributes real formative power to anything other than the gene. Again, in discussing the preformation-epigenesis dispute, he says:

> No preformed and complete structure preexisted anywhere; but the architectural plan for it was present in its very constituents. It can therefore come into being spontaneously and autonomously, without outside help and without the injection of additional information. The necessary information was present, but unexpressed, in the constituents. The epigenetic building of a structure is not a *creation;* it is a *revelation.* (1971, p. 7)

And, after another vivid description of development, he concludes, "The determining cause of the entire phenomenon, its source, is finally the genetic information represented by the sum of the polypeptide sequences, interpreted—or, to be more exact, screened—by the initial conditions" (p. 95). If by "initial conditions" he means only the intracellular environment when the molecular chain folds through the third dimension, he is excluding from this story all other nongenetic conditions, including substrates, intracellular machinery, factors influencing occurrence and rate of enzymatic action, as well as interactions at the tissue, organ, and organism levels that can, at best, be only partially explained by stereospecific interactions among molecules. Even if he includes these sources of specificity (and therefore of "information"), they evidently do not qualify as "determining causes" or "sources" of the phenomenon itself.

I dwell on Monod at length because he is a skilled and expressive writer; because it is difficult to question his credentials in molecular biology; because he has chosen to address issues beyond that field, such as ontogeny, evolution, and values; and because, given his standing as a scientist, he has unusual authority when he does so. He is thus in a powerful position to influence the thinking of his colleagues and of the general public, particularly because this conceptual ground is so well prepared.

In reviewing *Chance and Necessity,* Toulmin (1982, pp. 140–155) places Monod's thinking (and his rhetorical style) in the context of a French intellectual tradition that has been slow to relinquish progressivist ideas of evolution. Monod's emphasis on the role of the random in evolution and on the one-way flow of information from gene to protein is perhaps best understood in this light. Ravin (1977) points out that this latter notion of unidirectional transfer of information, Crick's "central dogma," is a denial of the inheritance of acquired characters. It is, however, unnecessary to deny nongenetic contributions to biological form in order to deny Lamarckian inheritance. As Toulmin points out in the review cited above, even randomness of variation is not essential to an anti-Lamarckian argument, though "decoupling" of variation from selection is.

One might argue that Monod is concerned not so much with the particulars of Lamarckian evolution (some of which Darwin did not himself reject) as with Lamarckianism as emblematic of an unscientific vitalism. Not being above using "vitalistic" and "preformationistic" as terms of abuse myself, I would point out two things: First, both are descriptive of particular kinds of attitudes and explanations, as well as of broad philosophical-theoretical positions, and they can coexist within theories and within persons. Second, any approach to biological processes that begins with inert raw materials requires a mindlike force to fashion this matter into a functioning animal-machine. This approach is at odds with what we know about physics and chemistry, including Monod's own findings about the interactions of complex molecules. The implications of this view of development as revelation will be elaborated in later chapters. At this point, however, lest it seem that I am singling out a scholar whose primary field is not development, and in fact one who has every reason to be particularly impressed with genetic functioning, let me point out that the views described are not at all peculiar to Monod, or to molecular biologists. They exemplify much thinking in biology and psychology. Indeed, if this were not so I would not be writing this book.

Decades ago, a sensitive and sophisticated observer of human development made similar statements about the causal primacy of the gene. In his much-quoted pronouncement, Arnold Gesell declared that environmental factors "support, inflect, and specify, but they do not engender the basic forms and sequences of ontogeny" (1954, p. 354; see discussion in Oyama, 1982). Oppenheim (1982) has argued that Gesell sometimes belied his understanding of development in his efforts to counteract excessive emphasis on the environment. This seems correct and important to me; it is often the case that we must temper our reading of documents with such historical perspective. As was the case with Monod's polemic for scientific Darwinism and against mystical Lamarckianism, however, it is important to distinguish justifiable implications from false ones. One of the problems with these grand oppositions is that, while they may fuel investigation and theoretical advance, by placing those activities in an erroneous context they make correct inference improbable. Opponents tend to pay more attention to refuting each other's claims than to examining the logical bases for those claims. By combatting environmental determinism with special formative, engendering genetic causes, Gesell only legitimized the underlying assumption that genetic and environmental causes could and should be distinguished in this way. He thus helped perpetuate the empiricist-nativist opposition, when, as Oppenheim and I point out, his approach potentially transcended it.

It is the structure of the argument, finally, that intrigues and troubles. Almost three hundred years ago, preformationists wondered whether the germ, the complete but inert form, was in the sperm or the egg. Because fertilization was necessary to activate development, there were, as for

Gesell and Monod, two necessary sets of interactants, neither sufficient to produce the organism. Both influences were granted power to affect the outcome (the parent who did not contribute the germ could still impart some qualities to the offspring). The *basic* form, however, was contributed by only the egg or only the sperm, depending on which camp one was in (Jacob, 1973, pp. 57–59).[5]

The assumption seems to be that change of the sort we see in developmental processes must have a single, fundamental source; since such change is ordered and directional, the problem becomes essentially that of the origin of form. Having defined an induced pattern as one that is "imposed . . . by the immediate environment," Bonner comments that the pattern resulting from embryological induction lies not in the inductor but "largely in the stimulated tissues" (1974, pp. 221, 249). Much of the history of embryology has been the chronicle of attempts to locate patterns in tissues or in chemicals (see Waddington, 1962, pp. 190–195, on patterns and prepatterns), just as that of the study of behavioral development has frequently been an exercise in pronouncing patterns to be inborn or learned. The problem, of course, is that patterns as such don't exist anywhere before they are realized—constructed by reciprocal selection (or coaction—McClearn and DeFries, 1973, p. 311; and Oyama, 1981) in ontogeny.

B. Determination and Commitment

Yund and Germeraad follow standard procedure in declaring that development results from a "highly coordinated program of gene activity. The genome itself is the program. It must contain all the information for its own regulation." They admit, though, that little is known about the "branching" of the developmental program governing gene expression in multicellular organisms (1980, p. 317). They go on to describe cellular determination as "commitment" to a "particular branch of the developmental program" and outline the methods of investigating the time of determination by looking at subsequent differentiation of cells in vitro or after transplantation. Interestingly enough, they remark that the more sophisticated the manipulation, the fewer are the cases of completely stable, irreversible commitment. (See also Alberts et al., 1983, p. 835, who state that the crucial characteristic of determination is not irreversibility but heritability, defined as "self-perpetuating change of internal character that distinguishes it [the cell] and its progeny from other cells in the embryo and that commits these progeny to a specialized course of development." Note that "heritability" here refers not to genetic changes but to enduring alterations in cellular processes that distinguish the cell and its progeny from other cells. See also Margulis, 1981, pp. 164, 177, 224.) So dependent is commitment on the method of investigation and the particular question asked that its strict definition as stable cell fate has been abandoned. In its place is a conception in which a "separate decision, regardless of its stability, occurs at each branch in the program." One can say only that a cell is on a particular path at a particular time, and investigate the mechanisms and relative stability of the decision. Its developmental potential, say the authors, narrows with each decision (Yund and Germeraad, 1980, pp. 317–319). Surely this is not consistent with the idea that the cell's genetic program contains "information for its own regulation," but rather with an indeterminate process in time, whose regulation depends on conditions. "Decisions" are not written in the nucleus but are made on the basis of developmental contingency. It is no wonder that determination becomes largely a matter of methodology, since "decision" is merely an anthropomorphic locution for a consequence. Individuals may decide to become this or that, and resolutely commit themselves to a course of action. Cells change states or not, depending on their competence, which may change, and their surroundings, which may also change. Their "commitment" must therefore be assessed with respect to those considerations.

Under some circumstances, a peculiar phenomenon known as transdetermination may be observed. Transplanted imaginal disc tissue from the larva of the fly *Drosophila,* for example, usually behaves in a very "programmed" (i.e., predictable) manner, apparently being determined or committed very early in development to give rise to certain adult structures. It will sometimes, however, differentiate into structures characteristic of *other* discs. This new determined state can be propagated by serial cloning. Furthermore, transdetermination may occur repeatedly, with successive transplantations. The authors note that the ability to transdetermine depends on the developmental state of the cells, especially age, and on original position in the imaginal disc (Yund and Germeraad, 1980, pp. 333–335). Once again we observe that investigators' descriptions of actual phenomena give the lie to their general pronouncements on the nature of development. It is not even the case that each developmental decision narrows the cell's potential, any more than an individual's does; in transdetermination, potential seems to shift and *increase* in scope (Alberts et al., 1983, pp. 833–840).

Conceptually the problem of determination is parallel to that of the sensitive period as fixer of developmental fate. The behavioral notion of the sensitive period is in fact derived from the embryological one of critical period and tissue determination (Oyama, 1979). Detailed investigation in both cases reveals a phenomenon that is not absolute and definitive, but complex and relative, one that is much more consistent with the model of development being elaborated here than with any fixed program.

A corrective for a person who tends to think too much in terms of potters molding clay or of computers printing out messages might be the idea of campers raising and stabilizing a tent pole by pulling in opposite directions. Stability is dynamic, clearly depends on both participants, and may be maintained to the extent that variation from one or both directions can be compensated for. Attributing the general outcome to one camper and trivial details to the other would falsify the process. I hasten to add that I don't consider this an adequate metaphor for ontogeny, but rather an illustration of a fairly simple point about causation: that it is multiple, interdependent, and complex. Even the potter, in fact, does not command absolutely. An artisan respects the qualities and limits of the material as much as he or she does his or her own; much of artistry, in fact, lies in just this respect for, and sensitivity to, the medium and the developing form. Finally, a program, to be useful, must be responsive to its data; outcomes are jointly determined.

It would seem that polemics do not justify arguments of the Monod-Gesell type, which, being characterized by their structure, predate these two scientists and can be found on the environment side as well. They do not adequately represent the analytic reality that the author is attempting to convey. In fact, by misattributing the orderliness of these processes, they diminish the impact of the argument.

Change, then, is best thought of not as the result of a dose of form and animation from some causal agent, but rather as a system alteration jointly determined by contemporary influences and by the state of the system, which state represents the synthesis of earlier interactions. The functions of the gene or of any other influence can be understood only in relation to the system in which they are involved. The biological relevance of any influence, and therefore the very "information" it conveys, is jointly determined, frequently in a statistically interactive, not an additive, manner, by that influence and the system state it influences.

In mice, many cell types contain the same androgen receptor protein, but different types of cells respond to androgen stimulation by activating different genes (Paigen, 1980). The *Drosophila engrailed* mutation influences sex combs, bristles, or vein patterns, depending on the appendage. The authors who cite this research observe that organs seem not to have their own "subplans" but to use general mechanisms in specific ways (Leighton and Loomis, 1980, p. xvi). The impact of sensory stimuli is a joint function of the stimuli and the sensing organism; the "effective stimulus" is defined by the organism that is affected by it. That one creature's sensory meat is another's poison, or that the same stimulus may have different effects on the same organism at different times, does not render stimulation causally irrelevant or merely permissive (as opposed to formative). The concepts of motivation, personality, and maturation have been developed in part to address such contingency of stimulus effects. Similarly the gene controls by being controlled or selected, sometimes by other parts of the genome, and even regulatory genes regulate by being regulated (Kolata, 1984). One of the prime questions of developmental biology, in fact, is that of differential gene activation.

Causation is endlessly interlocked, and the biological "meaning" of changes depends on the level of analysis and the state of the whole. This perspective may make it more difficult to say with confidence what constitutes a "whole" or a "system" in any given case, but since the material of life is neither structureless nor inert, there is no need for animistic forces; form and control are defined in life processes, not the other way around. (See also Waddington, 1972, p. III.)

In describing the nested feedback loops (which, by definition, involve reciprocal control) through which the environment influences developmental processes, Bonner declares the search for ultimate control factors to be futile, unless one conceives of ultimate control as that which

comes from the power gained by repeated, successive life cycles. Only in this way can one achieve a vast array of complex, interacting events.

Successive life cycles allow the accumulated information of millions of years to be used at a moment's notice. As far as control mechanisms go, there is no end: they go back to the beginning of life and are part of what we have called evolutionary development. The products of that control information are realized in each life cycle. (1974, pp. 156–157)

When Bonner asserts that life cycles may use "information" developed in previous cycles, he is referring primarily to cell products and structures already in place when a new sequence of ontogenetic differentiation begins. His discussion of nongenetic inheritance, in which "direct inheritance from one generation to the next is not restricted to the DNA of the genome, but many other substances and structures are built up from previous cell cycles" (p. 180), could readily be extended to include extracellular, even extraorganismic influences. The notion of life cycles as nested feedback loops (or at least nested sets of relationships) is a crucial one that deserves wider generalization to include not only pheromonal cycles in social insects, which Bonner considers to be the largest ontogenetic loop, but other essential ecological relationships as well, *even when they involve the inanimate world or other organisms.*

What we are moving toward is a conception of a developmental system, not as the reading off of a preexisting code, but as a complex of interacting influences, some inside the organism's skin, some external to it, and including its ecological niche in all its spatial and temporal aspects, *many of which are typically passed on in reproduction* either because they are in some way tied to the organism's (or its conspecifics') activities or characteristics or because they are stable features of the general environment. It is in this ontogenetic crucible that form appears and is transformed, not because it is immanent in some interactants and nourished by others, or because some interactants select from a range of forms present in others, but because any form is created by the precise activity of the system. Since even species-typical "programmed" form is not one but a near-infinite series in transition throughout the life cycle, each whole and functional in its own way, to refer to the type or the typical is also to refer to this series and the constant change that generates it.

If the genome, highly structured and integrated as it is, cannot by itself explain the products of ontogenetic change (the cognitive, planning function), can it at least be seen as the driving force (the causal, volitional,

energetic function) of such change? Much is written about the genes initiating, engendering, and originating, and the idea of diminutive chemical
engines powering biological processes is appealing. In fact, of course, a
gene initiates a sequence of events *only if one chooses to begin analysis at
that point;* it occupies no privileged energetic position outside the flux of
physical interactions that constitutes the natural (and the artificial) world.
A seed may remain dormant for years, and though plants frequently show
this kind of developmental passivity,[6] it is observed among animals as
well (Clutter, 1978). The seed "initiates" a period of growth when it is
triggered by some change. Genes affect biological processes because they
are reactive, and this reactivity is a prime characteristic of our world, at
all levels of analysis, from the subatomic through the social to the astro-
nomical. To describe biological processes as the product of exchanges of
energy, matter, and information (see Chapter 5), while consistent with the
temper of the times, is misleading in seeming to postulate a third, quasi-
physical force at work in the world. Both the initiation and the course of
biological change are functions of developmental systems, and there is no
evidence that our notions of matter and energy exchanges, themselves admittedly evolving, are inadequate to describe them. Adding information
to matter and energy is something like speaking of nations exchanging
dollars, yen, and profits. The third term belongs on a different level. Not
another form of currency, it describes a certain disposition and use of
currencies. Just as time or information can, under certain circumstances,
"be" money, matter and energy can sometimes "be" information.

It may be objected that this view of development as the result of changing and widely ramified systems is too complex and multiple, that its
boundaries are too indistinct, potentially extending to anything and everything in the universe. Yet this is just the kind of world scientists show us.
We can include only a few of its aspects in our models, and our choices
can always be shown wrong. In fact, when we attempt to provide artificial
conditions for proper development of captive animals, we are often forcefully reminded of the breadth and variety of linkages between biological
processes and extraorganismic factors. When feeding or reproductive "instincts" fail in captivity, we are prompted to search for missing factors in
the vital system, often without success. What is programmed, committed,
or determined, switched or triggered, depends on external considerations
that are as causally basic to the design of the phenomenon as internal fac-

tors, whether or not they are included in the design of the researcher, and however they are designated after the fact.

In living beings, no agent is needed to initiate sequences of change or to guide them to their proper goals. Matter, including living matter, is inherently reactive, and change, far from being an intrusion into some static natural order, is inevitable.

4 Variability and Ontogenetic Differentiation

If the homunculoid, cognitive-causal gene is not an adequate explanation for constancy of form or for ontogenetic change, what is its status with respect to variability, that is, differences among entities? One would think that if genes are used to explain developmental constancy, "something else" would have to account for variability. When, in fact, several phenotypes arise from the same genotype, differences are attributed to environmental variation. ("Phenotypic variation" is described by Bonner, 1974, p. 114, as "variation that is not inherited." In the present discussion, all it means is differences among phenotypes; no distinction between inherited and noninherited variation is implied.) According to the experimental model that holds some things constant, varies others, and accounts for variation in outcome by variation in input, this is analytically impeccable, *as long as differences are not confused with the phenomena themselves.* Natural examples of what one might call *phenotypic radiation* are found whenever clonal sets exist; conceptually it appears in the idea of norm of reaction, the set of phenotypes obtained when organisms with the same genotype develop in different environments. Monozygotic twins are a mainstay of genetic research on humans, as are inbred strains for research on other animals, because differences between them are attributable to conditions of rearing. Phenotypic variability in natural populations is sometimes assimilated to this idea, but only insofar as the usually erroneous assumption is made that genotypes are uniform throughout the population. (See Allen, 1979, for some comments on the history of the genotype-phenotype distinction.) A moment's reflection, however, reveals that ontogenetic differentiation, the appearance of increasing but integrated diversity in cells with the same genome, presents exactly the same situation: the development of different phenotypes from the same genotype, in what we could call *ontogenetic radiation.* Yet, as we saw, orthodox thought somehow explains this type of radiation by the *genes.* It is deeply inconsistent to explain the development of variable phenotypes from identical genetic "information" by that same genetic information when variability is generated within an organism's skin, but by the environment when variability is generated among organisms. This is nevertheless one basis for the traditional distinction between hereditary and acquired traits. Only the former, of course, are considered evolutionarily significant, because only they will be passed on, allowing evolution, including *phylogenetic radiation,* to occur.

A. Heredity and Heritability

If "hereditary" means that which is "passed on" *in some way or other,* then the statement that only hereditary traits will be passed on is tautologically true. But "hereditary" usually means "passed on in the genes," and since traits cannot literally be passed on in the genes or anyplace else, it is a mischievous tautology. Though we all know that there are no hooves or noses in the genes, the accepted formulation is that the genes that are literally passed on make hooves and noses in ontogenesis. But if, as has been argued in preceding chapters, genes do not create traits according to a plan written in their very structure, even by operating on conveniently available "raw materials," if phenotypic characteristics arise only when sufficient interactants are present in the proper place and at the proper time, and if all these factors are therefore given comparable causal and formative significance, *then defining heredity as the passing on of all developmental conditions, in whatever manner,* is preferable to defining it by genetic information. This does not require any distinction between hereditary and acquired traits, or even between mostly hereditary and mostly acquired ones; all it requires is some degree of association of developmental influences. It is the relative unreliability of association that distinguishes the unusual phenotype, which does not reappear in successive generations, from the common one that arises again and again from the species pool of developmental means.

Since the sophisticated, except in unguarded moments, do not usually speak of inherited traits, but of inherited or heritable variability, there is something of a straw man in the above argument. The switch to the language of heritability, however, unless extremely tightly constrained, proliferates confusion rather than dispelling it. Heritability, understood as the proportion of phenotypic variance attributable to genetic variation under controlled conditions, is not a characteristic of traits but of relationships in a population observed in a particular setting. These relationships are expressed in numbers, which depend on the precise levels of genetic and environmental variables examined and the selection and operationalization of the dependent variable(s). Heritability, that is, is an attribute not of variants but of their statistical descriptions (variance). These descriptions are as dependent on the research design as they are on the traits themselves. Thus, how much variability is accounted for by one or another factor depends on what analytical question is being asked, when and across what range of phenomena. By the same token, the extent to which a character responds to selective breeding, another way heritability is measured, depends very much on the precise manner in which selective pressure is brought to bear. What induces variation, and how much variation results, depends on what the character is and what the conditions are. We will return to these issues when we discuss the developmental contingency of information. Heritability coefficients, in any case, because they refer not only to variation in genotype but to everything that varied (was passed on) with it, only beg the question of what is passed on in evolution. All too often heritability estimates obtained in one setting are used to infer something about an evolutionary process that occurred under conditions, and with respect to a gene pool, about which little is known. Nor do such estimates tell us anything about development.

All three kinds of radiation—(1) ontogenetic (differentiation), (2) phenotypic (variation among individuals with the same genotype, norm of reaction) and (3) evolutionary (phylogenetic divergence)—share the qualities of constancy, change, and variability. The last two kinds of radiation, in addition, assume the first, in that variant phenotypes must develop and phylogenetic variability is diversity in ways of developing. To put it another way, phenotypic radiation is differences in ways of changing, and evolutionary radiation is changes in ways of changing, including changes both in gene pool and in effective niche. Though ontogenetic radiation is traditionally explained by the genes, and phenotypic radiation by the environment, they are not even unequivocally different processes.[1] A succession of life cycles, phylogeny is not a determinant of those life cycles, except in the sense that the characteristics of a cycle help constrain the types of novelty that can arise and be functional in the next. Far from being an efficient or mechanical cause of ontogeny, either literally or in the sense of shaping genetic codes that direct ontogeny, a phylogeny is simply descriptive of progressive variations in ontogenesis.

Though evolution tends to be seen as the consequence of "environmental pressure" or "shaping by the environment," or even "molecular instruction" by which the genetic code "learns" to make improved responses (Eigen, 1967, p. 141), such formulations are inadequate for the same reasons that genetic or environmental controls are inadequate to explain ontogeny: they attempt to account for the qualities of outcomes by allocating formative power to one of several jointly necessary and singly insufficient interactants. Environmental pressure can be defined only with respect to a particular developmental system; the evolutionary significance of that pressure can emerge only from interaction, just as the ontogenetic "message" encoded in a genome is only defined (not revealed) in development. Lewontin observes that ecological niches "can be specified only by the organisms that occupy them." (Notice the similarity of this argument to that presented above on effective stimuli.) Organisms transform, create, and choose niches; "although natural selection may be adapting the organism to a particular set of environmental circumstances, the evolution of the organism itself changes those circumstances" (1978, p. 215). Natural selection must be understood as an interactive process whose very constraints and causes emerge as it functions, as they do in a developmental process. While it is true that ontogeny is generally predictable in ways that phylogeny is not, this is because in species-typical ontogeny the typical developmental interactants are by definition predictably present and closely integrated, physiologically and ecologically; the concept of the developmental system is based on the indispensability of both gene pool and species environment to heredity. Phylogenetic sequences are not repeatable in this manner. It is interesting, however, that kinds of ecological relationships do recur in the phenomenon of evolutionary convergence, even though they may be enacted by very different species. In convergence, similar traits evolve in similar ecological circumstances, but by different phylogenetic routes. The tendency of an ecosystem to restore itself after some disturbances is also reminiscent of embryological regulation (understood not as events regulated by a directive agency but as tendency to reach a more or less normal state despite some perturbation). May (1978, p. 161) gives descriptions of both phenomena and remarks, "the forces shaping natural selection (N.B., natural selection as a process that can be influenced, not as a prime mover) among individuals involve all manner of biological interactions with other species," and "constellations of species can be viewed as evolving together within a conventional Darwinian framework."

The comparison (and it is a limited one) between ontogeny and phylogeny can be made, however, only if the idea of ontogeny as genetically directed development is given up. The present conception of ontogeny as inherently contingent, rather than preprogrammed, eliminates the specter of orthogenesis that has troubled previous attempts to liken evolution to development. Jacob describes evolution as tinkering, or "bricolage." Rather than working from a "preconceived plan" as an engineer would, it improvises and makes do, in a "series of contingent events" (1982, pp. 33–37). Instead of distinguishing evolution on these grounds from development as he does, I am suggesting that *both* processes show the contingent quality of tinkering, in the sense not of randomness or disorder but rather of subtle and opportunistic dependence on particular conditions and materials. This accounts for both their vulnerability and their power, as well as their "ingenuity." Rather than the directedness of planned activity, it is such inspired tinkering that characterizes life processes, the marvelous results notwithstanding. In the case of normal development, however, the scraps and bits of twine are all at hand, and the qualities, needs, and limitations that must be respected are also part of the species system; under such conditions a good tinkerer might turn out a succession of very similar objects.

Enter the exasperated reader: "It's all very well to make these distinctions between variants and variance and to quibble about the relationship between ontogeny and phylogeny, but the bottom line is that, if I cut off a mouse's tail, its offspring will have normal mouse-length tails, because the normal tail is inherited, heritable, phylogenetically encoded, genetically predisposed, or I-don't-care-what, and the stump is not."

This apparently unequivocal objection deserves some consideration. Surely the knife that chops off the tail is an "environmental" cause, and it in no way alters the mouse's germ cells. But it is also not a reliable aspect of the mouse's world. If short tails were conducive to survival and successful reproduction, perhaps because they made the mouse harder to catch, and if tails were reliably chopped off, short tails would be species-typical, functional, and stable across generations. (Even if they were not beneficial, they would reappear as long as they did not seriously reduce reproductive effectiveness. If short tails were detrimental, one would expect that the creatures either would become extinct or would evolve some chop-evading or chop-neutralizing tactics.) If, in the absence of ubiquitous and indiscriminate tail choppers, (1) some mice were more likely than others to invoke the choppers' wrath (or, rather, to win their favor), then whatever character, structural or functional, made that difference would also be beneficial, as would (2) a change in morphogenetic dynamics that resulted in shorter tails, (3) maternal care that involved biting off the tip of the tail, perhaps at birth, and (4) a habit of folding the tail against the body, thereby functionally shortening it.

Each of these "solutions," of course, would be truly beneficial only if it did not exact too great a cost in other areas of the mouse's life and if the selective significance of short tails remained unaltered by these associated changes. Each, in addition, could initially be accomplished with no change in the gene pool, though consistent selection could well lead to such change. (1) Mice living near human settlements, for example, might be more likely to experience tail amputation; (2) local conditions might alter maternal nutrition or endocrine functioning in such a way that caudal morphogeny would be altered in offspring; (3) local conditions might alter maternal behavior either directly or indirectly—by changing the infant's behavior, appearance, or odor, for example (see P. P. G. Bateson, 1982, on the reduction in rat-pup distress calls in the laboratory, as opposed to the wild, and the resulting decrease in maternal care); or (4) sequences of experiences might reliably occur, perhaps through alterations in terrain or in climate, such that the habit of tail tucking would arise predictably. A series of alternatives could be imagined for each of these possibilities. Developmental psychology, developmental biology, ethology, and comparative psychology supply an ample number of examples of developmental contingencies to demonstrate that these ontogenetic options, though simplified and fanciful, are not fully as outrageously implausible as they may first appear.[2]

Perpetuation of short tails, then, regardless of their way of coming into being, depends, quite simply, on the reliable association of relevant genotype and conditions, what Gottlieb calls the "entire developmental mani-

fold" (1975, p. 684). The association may be due to their being interactants in embryological processes, in which case tails will be short from
birth, and/or to their being part of species-typical networks of interactions
of organisms with themselves, conspecifics, members of other species,
or the inanimate world. Any of these developmental outcomes could be
altered by genetic or environmental changes, or both; this would not make
the trait more or less genetic or environmental, more or less preprogrammed, just different. A given genotype has associated with it a whole (partially) nested set of environmental conditions, and when genotype shifts,
so may many of those conditions, some changes occurring as a fairly direct result of different gene products or patterns of transcription, some as
a result of different reproductive environment, some as a result of different sensory capacities, and so on. Conversely, a shift in conditions may
make itself phenotypically evident in altered gene transcription resulting
from a change in substrate level or developmental timing. What counts,
from an evolutionary point of view, is the reliability of the character, that
is, the likelihood that adequate interactants will be reassembled in each
generation (reproductive success). If differences in genes are related, in
whatever way, to reproductive success, and if this relationship holds over
many generations, gene pool constitution may shift; if not, modal phenotype may still shift in the absence of substantial genetic changes. It would
seem that such a shift could be seen as evolutionarily important insofar
as it influenced the nature and distribution of phenotypes and of selective
pressures on the population, and thus the kinds of genetic changes that
were likely to be viable or beneficial in the new developmental system.
As P. P. G. Bateson remarks, the precise manner of a character's coming
into being is irrelevant to natural selection (1982; see also Dunbar, 1982).
It is possible, of course, that certain ways of developing themselves may
have a selective advantage over other ways. Adaptive advantage, furthermore, is not the only factor shaping gene pools, but that, as they say, is
another story.

Just as biological regularity is a joint product of "control" by genes and
conditions, so is any particular variation. This is why statements about
sources of differences or lack of differences must always specify the conditions of the comparison. And just as a correlation between genotype
and fitness is uninformative about developmental dynamics, details of de-

velopmental dynamics of a character are apt to be uninformative about its evolutionary significance.

The ambiguity of such terms as "genetically programmed" derives from the fact that they are used to make statements both about evolutionary history and about ontogenetic mechanisms (often simultaneously, since the two are assumed to imply each other). The ambiguity is multiplied because the historical statement may be about universality of the character in an ancestral population, its universality in a contemporary one, or about the selective consequences of its distribution in one or more such populations. Similarly, the developmental implications may embrace any or all of the traditional meanings of the innate. As students of evolution and of ontogeny refine distinctions in these areas, the need for an unambiguous vocabulary becomes more and more urgent. Describing the results of cellular differentiation in ontogeny as genetically determined (*despite* environmental determination of differences), like describing the different phenotypes that may arise from the same genotype as environmentally shaped (*because of* environmental determination of differences), misrepresents ontogeny while it dilutes the utility of the notion of determination of variance. This is also true of descriptions of evolution as environmentally directed, despite the mutual selection of niche and organism, and despite the fact that natural selection is the cumulative result of those interactions, not their cause.

It should be clear by this point that it is useless to take refuge in genetic predispositions or potential, though it is an exceedingly common recourse. Bonner, for example, in describing a sensitive period for development of egg polarity in an alga, speaks of organisms that "inherit an ability to respond to the environment during their development" (1974, p. 221). While it is undoubtedly true that the alga "inherits" this developmental competence in the sense that the competence is species-typical and therefore reliably correlated with genotype, it also "inherits" the stimuli by which the competence is defined, and its developmental sensitivity is a joint function of conditions and genes in the same way other phenotypic characters are. The idea of inheriting an environment may make more intuitive sense when the environment is part of a maternal reproductive system, but the conceptual differences between this and the broader notion of a developmental system are minimal, particularly when the maternal genome is not identical with that of the offspring. Only part of development, furthermore, occurs within the egg or within the mother's body. One could argue that the chemical and energetic characteristics of the surround are part of the algal developmental system; this is, in fact, what I do argue. If "inherit" is being used in the source-of-differences sense, its informativeness is related to the specific analytic context in the same way these terms always are. If, on the other hand, it is being used to indicate a kind of developmental causation, it is inaccurate for the same reason all such uses are. Disembodying characters by making them potential, that is, does not solve or even attenuate the conceptual problems inherent in all nature-nurture partitionings.

Acknowledging the integrated nature of inheritance, that what is quite literally passed on or made available in reproduction is a genome and a segment of the world (Montague, 1966), though the boundaries and constitution of that segment change with time, allows us to describe such events as the alga's interaction with its surround as maturational, that is, part of the species-typical developmental system. This is so even though a particular external stimulus (in this case light and various chemicals) is involved. This way of thinking about inheritance seems quite congenial with Bonner's general orientation, given his appreciation of the multiplicity of relations that make up a life cycle and the impossibility of excluding extragenomic but species-typical factors from their description. This, in fact, is precisely the meaning of "inherit" that I read in his account, and this meaning of inheritance does not depend on the presence of heritable variation in photosensitivity, for instance, in any particular population of these organisms or on a conception of development that is explicable by internal direction alone.

B. Genes for Traits, Nails for Battles

Dawkins (1976, p. 66) believes that consistent substitution of the language of genetic differences for that of genetic traits eliminates the problems that developmentalists have persistently found with sociobiological accounts. Though this rigor of expression is important, it does not confer license to proliferate genes for everything imaginable. If he is to be faithful to his own descriptions of ontogeny and of the role of genes in making a difference between two processes that themselves require much complex interaction to develop, he must, and does, admit that a gene is "for" a given difference only when everything else, including other genes, remains the same, or is at least functionally equivalent.

It may be that the third nail on horse Dobbin's left hind hoof was loose, and that the loss of that shoe lost the horse and, ultimately, the battle. But for that nail to play the same role repeatedly (and thus to be a nail "for" losing battles), a great many things would have to be present in the same configuration, as in fact they tend to be in embryogeny. We cannot have a gene for brown rather than blue eyes unless we are assured of having eyes (and nerves and skins and everything else required to make an organism with eyes). Similarly we cannot have a gene for cheating in particular encounters unless everything else necessary to constitute those encounters is undisturbed or equivalent, and here the requirements go far beyond physiological integrity of the individual, embracing all sorts of behavioral interactions as well. What makes a given action an instance of cheating, in fact, depends less on the details of the action than it does on the role it plays in the entire transaction (to say nothing of intention and belief, without which the term is presumably used metaphorically). For a gene to be "for" a given social act, then, in even a small proportion of the organisms in which the gene appears, there must be enough developmental-behavioral consistency for all concerned to reconstitute the essentials of the situation in at least that many cases. Though Dawkins rightly says that particulate inheritance does not require particulate embryology (1982, p. 116), we must exercise considerable restraint to avoid confusing the two, especially given their firm association in many minds. A rhetorical style that persistently projects intention and affect onto large molecules, I suggest, makes it more, not less, difficult to distinguish particulate inheritance (gene transmission) from particulate embryology (trait transmission) because it encourages us to think of genes, which may affect development in certain ways, conditions permitting, the same way we think about persons, who may search for ways to attain their chosen ends regardless of conditions.

Developmental issues cannot be divorced from speculations about evolution. Though it is correct that natural selection is in some sense blind to developmental mechanics, this does not mean that we may postulate genes for anything we please and trust the mechanics to be feasible. It is only by understanding development that we can know how external and internal conditions are linked and learn the range of those conditions over which a given gene will have a given effect. Furthermore, it is only by investigating relations within the developmental system that we will learn the extent to which the conditions for developing that effect are associated with the conditions in which it will be advantageous. This is where the developmentalist critics of the adaptationist program have an important point to make. Being uninterested in developmental dynamics, that is, does not make everything possible, and even what is useful is useful only in a particular context.

The more reliable and well integrated a developmental system is, whether it involves gestation and rearing, interspecies interactions (including dispersion of seeds by animals who ingest them or carry them on their bodies and many other plant-animal interdependencies), or other ecological relationships, the more spontaneous and inevitable—the more "biological"—ontogeny will appear, and the more uniform the effects of any given genetic or environmental factor will be. The more unpredictable the system, the less programmed it will seem.

Biological constancy, change, and variability can be specified only by describing these developmental systems, whether we are speaking of ontogeny or phylogeny. This is something everybody agrees with, in some way or another, just as nearly everybody is an "interactionist," which makes it both difficult and tiresome to keep pointing out that adoption of correct terms is no guarantee of coherent use of ideas, even though it ensures superficial agreement and therefore the illusion of conceptual mastery. Why, then, do we continue proliferating ever more subtle, and thus ever more difficult to recognize, versions of the old distinctions? For one thing, our most fundamental convictions about life processes have not seemed to keep step with our growing factual knowledge (Weiss, 1969, p. 34). We create and observe *differences,* but we wish to understand *phenomena.* In attempting to explain phenomena, we use the tools that served us so well in investigating those differences, and we fall into the fatal gap between analysis and synthesis, partitioning characters the way we partition variance and making pronouncements about traits and potential on the basis of F ratios. In a later chapter we will look more closely at the factors that seem to give rise to some people's continuing dissatisfaction with truly interactive views of development, and that therefore dispose them to adopt new ways of expressing their belief that form preexists its instantiations. What is just as significant in shaping our present vocabulary, though, is a set of powerful and evocative metaphors coming from molecular biology, information theory, and computer science. They are doubly attractive because they are not only consistent with these old beliefs, but validate them with biological and mathematical high tech.

5 Variations on a Theme:

Cognitive Metaphors and the Homunculoid Gene

In discussing variants of the homunculoid, cognitive-causal gene, I will group them for efficiency and will not cite specific authors for the most widely used ones. The groups are largely arbitrary, however, and any number of other variants and classifications could have been employed. That there is a great deal of common conceptual ground and overlap in usage is, in fact, one of the main points of this chapter. Yet a certain amount of detailed scrutiny of specific versions is justified because each seems, by invoking different images or processes, to offer a special intellectual advantage. It should be emphasized that these terms often form a mix-and-match pool, with distinctions made in one place dissolving in another, sometimes in the same discussion, and contradictory uses coexisting. In short, the same kind of conceptual disorder found in earlier versions of the nature-nurture distinction is found in the present vocabulary, even though this one claims to offer definitive resolution of the old dichotomy.

A. Genetic Blueprints and Plans

Discussed elsewhere (Hinde, 1968; Lehrman, 1970; Oyama, 1982), the blueprint metaphor does not require extensive treatment. Though a plan implies action, it does not itself act, so if the genes are a blueprint, something else is the contractor-construction worker. Though blueprints are usually contrasted with building materials, the genes are quite easily conceptualized as templates for building tools and materials; once so utilized, of course, they enter the developmental process and influence its course. The point of the blueprint analogy, though, does not seem to be to *illuminate* developmental processes, but rather to *assume* them and, in celebrating their regularity, to impute cognitive functions to the genes. How these functions are exercised is left unclear in this type of metaphor, except that the genetic plan is seen in some peculiar way to carry itself out, generating all the necessary steps in the necessary sequence. No light is shed on multiple developmental possibilities, species-typical or atypical.

B. Images and Knowledge

More idiosyncratic than the plan or blueprint, images and knowledge in the genes are described by Lorenz (1977) in an explicitly cognitive and economic treatment of life processes. Increases in knowledge about the world lead to economic (adaptive) advantages, which "then exert the selective pressure which causes the mechanisms that acquire and store information to develop further." In this scheme the "higher" organisms are those that store the most information. Knowledge is gained by the genome, which produces in the organism, by morphogeny, an image of the world, like a negative or a plaster cast; fish fins "reflect the hydrodynamic properties of water" for example, and human behavior, when it is adapted, is an "image of the environment." While the genome learns only from successes, Lorenz continues, man learns from both successes and failures, but both compare ideas or hypotheses with the world and acquire information from the degree of fit (pp. 6, 23–27). The examples make clear that images, casts, knowledge, and information all refer to adaptive fit between organism and niche, which Lorenz described with such perceptiveness and enthusiasm for decades. Far from addressing the problem of ontogeny, these emphases on well-integrated ecological relationships assume it, as the manner in which knowledge gained by the genome in evolution is expressed as "innate information" in the organism. By this latter term, Lorenz intends not polypeptide sequences, but a fairly everyday sense of knowledge, as in his example of the tick's innate information on what kinds of creature to bite (those with the proper smell and temperature [1977, p. 55]). Needless to say, however such well-adapted behavior is conceptualized, it is observed only in phenotypes, not in genotypes. Ticks may or may not be thought of as "having" minds, but I believe only minds can "have" knowledge.[1]

C. Symbols and Hypotheses

Apparently dissatisfied with simple and common cognitive metaphors, some authors elaborate them in an attempt to fit them to more complex views of biology. Pattee characterizes the genotype as a "simplified symbolic description" of the organism (1972, p. 248). He describes as the scope of theoretical biology "to explain the origin and operation (including the reliability and persistence) of the hierarchical constraints which harness matter to perform coherent functions." Except for the separation of constraints from matter that must be harnessed, this seems both reasonable and accurate. (In fact, later, on p. 250, he describes a constraint in physics as having a "distinct physical embodiment in the form of a structure.") He states that just as any description implies a language structure to make it meaningful, so molecules require a system of constraints in order to function as a message.

> We are taught more and more to accept the genetic instructions as nothing but ordinary macromolecules, and to forget the integrated constraints that endow what otherwise would indeed be ordinary molecules with their symbolic properties. . . . It is not the structure of molecules themselves, but the internal, self-interpretation of their structure as symbols that is the basis of life. And this interpretation is not a property of a single molecule, which is only a symbol vehicle, but a consequence of a coherent set of constraints with which they interact. (p. 249)

Assuming the "self" that performs the "self-interpretation" is not the molecules but the whole organism, Pattee appears to be making some important points about the explanatory insufficiency of DNA and about the relativity of the biological meaning of the molecules and their products to the physiological processes into which they enter. Whether these points are clarified or obscured by ever more elaborate metaphors, however, by which molecules gain symbolic properties in quasi-linguistic structures, is another question. If I am correct in interpreting Pattee's constraints as biological structures, processes, and their limits, he is moving in the correct direction: that is, toward the functioning phenotype-in-context, which not only is organized in part by the genes but directs and lends coherence to their activity. To describe the genome as a symbolic description, however, hinders this movement by reverting to the idea of form of the phenotype in the genotype.

Expressing similar concerns to Pattee's, Goodwin (1972) suggests that, although some people claim that genes specify embryological *processes,* it is better to see the genes as *hypotheses,* because a "hypothesis is always considerably less than a total set of instructions for its interpretation and test." Just as a verbal hypothesis presupposes much knowledge to be interpretable, so "hereditary material speaks to a competent hearer, and assumes that it can interpret and test the hypotheses it is presenting in some manner which will establish whether or not they have meaning." He states that genetic information, being in symbolic form, cannot interact with the world without being translated into the "language" of protein and goes on to describe hypotheses in the chromosomes of the egg "about the appropriate response of a cell in the ectoderm which is required to transform into part of the lens of the eye when it encounters a cell in the underlying optic cup" (pp. 269–270).

The competence that allows ectoderm to respond to induction by the optic cup arises only through many embryological events. In what sense does the hypothesis about induction exist in the nucleus of an *endo*dermal cell, or of the zygote, or, for that matter, of a cell in any tissue not competent to form a lens, including ectoderm at the wrong time? If it exists at all, doesn't it exist in just those cells that are inducible? When cells respond, furthermore, not as individuals but only as parts of ensembles, shouldn't the selective responsiveness be located in the tissue rather than in the constituent cells?

Where genetic hypotheses seem to be used to explain ontogeny in the preceding example, Goodwin appears in another to equate hypotheses with general adaptedness, in much the same way that Lorenz uses images and knowledge. He describes a bacterium whose genome includes the lactose operon. It carries in its genes the "hypothesis that lactose is likely to be encountered in its world and that this substance can be transformed in a particular way to provide energy." This is interesting because the "information" is about something outside the organism's skin, something that may or may not be present. Do our genes contain hypotheses about every stimulus to which we may respond? If so, doesn't this lead to inscribing all developmental contingencies, from the common to the rare, from the useful to the lethal, in the genes? If not, how would one distinguish between events about which the genes carried hypotheses and those about which no such predictions existed? If hypotheses exist only about adaptive or species-typical interactions, then nonadaptive or unusual ones must occur without benefit of such entities. How, then, do they occur at all? Must we postulate a distinction between improvised and planned gene activity? Wouldn't it eventually resolve to one between the acquired and the innate, thus entrapping us in circularity?

Goodwin says "the view of the organism as an hypothesis-generating and testing system is potentially very powerful" (1972, p. 267). Here he comes close to placing hypotheses in the phenotype. These might reasonably be conceptualized as ontogenetic and behavioral competences, or, more straightforwardly, current states of the organism. Phenotypes don't, in the usual sense, generate their own genotypes (though they do generate successive "effective genotypes" as genes are activated and repressed in different tissues at different times, and though changes in genetic material seem to be important in certain aspects of the development of certain organisms). Later in the paper, however, he notes that genuine hypotheses "at the level of mind" are subject to continuous alterations by experience, whereas genetic ones can be altered only from generation to generation (pp. 270–271). An organism, or better, gene pool or species as source of successive hypotheses, is thus given explicitly mindlike properties, but the hypothesis is static for the organism's lifetime and seems firmly embedded in the genotype, not the phenotype, whose competence is in transformation much of the time. Two organisms with identical genetic material may eventually develop different "hypotheses" about the world (a worker and a queen ant, for example), but Goodwin's metaphors do not seem equal to the task of expressing these facts. Like Pattee, he appears dissatisfied with the genes as full explanation of development and is cognizant of the complexity necessary for gene transcription to have ontogenetic efficacy. Yet he remains wedded to the cognitive gene and therefore misses making his point. In the same way that replacing traits in the gene with potentials solves no problems, so blurring the outlines of the genetic code by making it an incomplete description or only a conjecture merely makes the imagery more obscure without reducing the conceptual problems that motivated these exercises in metaphorical biology in the first place.

In the years following the article discussed above, Goodwin continued his questioning of traditional accounts of ontogeny and dropped these metaphors in favor of a less "genocentric" structuralist, constructivist description of development (personal communication, June 1984). He is a persistent critic of narrow selectionist views and has pursued the goal of articulating developmental with evolutionary biology. His 1972 ideas are included here because they show an important trend in the thinking of certain progressive developmentalists; just how progressive they were is evident when one realizes that these decades-old ideas are still substantially ahead of many contemporary treatments.

D. Rules, Instructions, and Programs

The genome as constituting rules, instructions, or a program, either in the sense of a plan or in the sense of a computer program, is so common a notion as not to seem metaphoric at all. Routine introductions to treatments of development present any one or a combination of these terms as the solution to the preformation-epigenesis problem. Yet, just as these molecules can be thought of as containing organs and behavior only by a prodigious stretch of the literary imagination, so is it difficult to understand how they could contain statements and commands. Yet, perhaps because rules imply activity rather than objects, these locutions raise few eyebrows. Several versions warrant comment.

Given the charges of genetic determinism that have been leveled against sociobiologists in general and E. O. Wilson in particular, it is interesting to note his adoption, with his collaborator, Charles Lumsden, of the term "epigenetic rule." They begin with an image of a naive human confronted with an array of cultural units, "culturgens," from which he or she makes a selection (Lumsden and Wilson, 1981, pp. 7–9). There are many difficulties with this representation of persons and culture, but it is the notion of the "innate epigenetic rule" that is of interest here. These rules bias culturgen choice, so that not all items are equally likely to be selected; the degree of bias is the degree of genetic transmission of culture (p. 23) and is related by the authors to "canalization" (pp. 242–243).[2] How this developmental bias works, however, is not made clear. The innate rules for humans, in fact, seem to be a set of candidates for species-typical perceptual, motivational, and motor patterns, some more speculative than others (color classification, sibling incest avoidance, facial expression, and mother-infant bonding, for example [p. 357]); they are more or less reliably observed products of human development, and from their postulated status as universals, epigenetic rules are inferred. Psychological, social, or cultural universals, then, are the result of such rules. The structural and physiological characteristics that underlie such traits are phenotypic, of course, and to the extent that they are typical of humans they are as much a function of shared environment as they are of shared genes. (Waddington's notion of canalization, in fact, was more complex than this; developmental systems tended, in his view, to maintain stability of outcome despite considerable variation in genes or environment.) Yet it is in the epigenetic rules that the "ultimate evolutionary goals of the mind" reside, and in one passage the rules seem themselves to live in the limbic system "rather than in the more purely cognitive and ratiocinating portions of the mind." This "onboard computer" in the brain is phenotypic, but the authors make clear—when they describe the causal sequence they are presenting as running from the genes to epigenesis to individual behavior to culture, with epigenetic rules serving as the "molecular units" of learning—that this is essentially a matter of translation of preexisting form (pp. 343–348). In fact, elsewhere in the book the authors show considerable comfort with the idea of "knowledge structures" that are "inherited as hardwired processes within the brain," as examples of "purely genetic transmission." Though they declare that this is rare in humans, whose behavior is based mainly on learned "knowledge structures," they clearly have no conceptual argument with inherited structures, and indeed, if epigenetic rules are the building blocks of learned structure, the distinction between inherited and learned structures is not clear (p. 255).

We see, then, that far from illuminating ontogeny, the Lumsden-Wilson epigenetic rule is identified by, and with, its allegedly universal products and is just another locution for the ghost in the ghost in the machine. As is commonly the case in such treatments, we are told that we ignore the rules and contemplate departing from the normal type at our peril (pp. 358–359). The classic circle is closed when epigenetic rules are first inferred from frequency of phenotype, then invoked to explain development of the phenotype. Though it may be futile to argue against all such circles, at least we may demand well-constructed ones. To identify our humanity with one of its ontogenetic sources is to endow adaptedness with a static quality that does justice neither to ontogeny nor to evolution.

Less restricted to particular authors than the preceding term, genetic instructions and programs are widely used. Ernst Mayr, one of the authors of the synthetic theory of evolution, defines a program as coded information controlling a process or behavior and leading it toward a goal. It is the result of natural selection, is material, preexists initiation of the teleonomic process, and is therefore "consistent with" causal explanation (1976, pp. 390–394). Concerned less with describing development than with accounting for purposeful behavior, Mayr's treatment is nonetheless interesting for the present discussion. Though he claims that the innate-acquired dichotomy is a false one because the innate refers to the genotype and the acquired to the phenotype, he maintains the dichotomy in his distinction between closed and open programs. The closed program, he says, allows no "appreciable modifications" during "translation into the phenotype"; "nothing can be inserted in it through experience," while an open one allows for "additional input during the lifespan" (1976, pp. 695–696). Such formulations, increasingly common, are an attempt to deal with the obvious contradictions that arose when the old innate-acquired opposition confronted the species-typicality of much learning. Yet the notion of "input" is still largely restricted to the environment as source of individual variation ("modification"), and workers continue to attribute species characteristics, and all the ontogenetic modification they imply, to the genes.

A second noteworthy aspect of Mayr's discussion of programs is that he, like Lumsden and Wilson, sometimes allows them phenotypic status. Behavior, he says, is actually controlled not by the genotype but by "behavior programs in the nervous system that resulted from the translation of the original genetic program"; when new information is inserted into an open program, it goes not into the genetic one, but into the translated one in the nervous system (1976, p. 696). As was the case with the epigenetic rule, the "program" in the genes can affect behavior only by being a part of phenotypic form and function. Forms and functions may indeed be species-typical, but *there is no developmental system that will generate them without the "insertion of information" from environmental exchange, and there is no atypical form or function that does not involve the same kind of reciprocally selective interaction between genetic constitution and developmental conditions.*[3]

The point here is simply that the characteristics of a system, including the "genetic system," determine, constrain, and control its functioning; no additional entity in the form of a program need be invoked. In fact, insofar as a separate plan is invoked, it mystifies the developmental process and encourages us to see variation as an intrusion.

Miller, Galanter, and Pribram define a Plan as "any hierarchical process in the organism that can control the order in which a sequence of operations is to be performed" (1960, p. 16), a definition that fits well with the concepts being developed here, insofar as a plan is a process rather than an object. Mayr even suggests that a program may be nothing more than the characteristics of the organism when he points out that, whereas the program in a computer is separate from the "executive machinery," the program of a loaded die is part of the machinery (1976, p. 394). In a personal communication, however (April 1983), he denies that *fair* dice have a program because nothing then determines the outcome. One could argue, though, that the "program" is for a random selection of one of six faces, and that this is quite different from the set of outcomes possible from a die of another geometric form or from a loaded one. What is seen as programmed depends on what comparison is being made, and therefore, what is meant by determination of an outcome.

In *The Growth of Biological Thought,* Mayr traces some of the links between the natural theological tradition of reading nature as a text and evolutionary biology, acknowledging that natural selection became in some sense a modernized hand of God. In describing the historical replacement of the Aristotelian psyche by the Christian soul, he comments that Descartes, by viewing the body as inanimate matter, condemned animals to soullessness and man to dualism. Molecular biologists, on the other hand, seem to have redefined the Aristotelian soul. Mayr sees the genetic program, a teleonomic formative principle organizing "raw matter," as a kind of Aristotelian essentialism fundamentally different from Plato's because it requires no outside force to explain complexity and purposiveness (1982, pp. 87–106). I am not able to say whether Aristotle's *eidos* really refers to the form of a thing, or to its forming, or both. I do think, though, that this ambiguity about whether the ultimate causes so popular in evolutionary biology refer to *functions* with a phylogenetic history or to macromolecular *agents* that guide the construction of forms to fulfill those functions is forever threatening the clarity and therefore the usefulness of the ultimate (what for)–proximate (how) distinction. It frequently collapses in on itself in the course of being employed, and past and present come to be seen as alternative developmental causes. Ultimate causes, that is, become not only contemporary, proximate causes (genetic), but the primary, initiating, formative ones at that.

Earlier in the same book, Mayr goes over the notion of programmed teleonomic activity that has so influenced scientist and layman alike (1982, pp. 45–51). Used to explain goal-directedness, it is distinguished from teleomatic processes (a falling rock, for instance), in which ends are reached by physical law, which he identifies with Aristotelian necessity. Though he quite rightly criticizes the reification of processes, like "life" and "mind" (pp. 74–75), it seems to me that he sometimes fails to heed his own good advice (who of us does not?), turning regular, organized ontogenetic, physiological, and behavioral processes into things, programs, that are then used to explain the processes. In doing so, he re-creates the conceptual and physical machinery for generating and regenerating the species essence he has shown to be antithetical to the modern evolutionary synthesis. In addition, by making a distinction between programmed teleonomic processes that run as they do because the program contains a goal, and teleomatic ones that run by mere physical necessity, he invites queries about why teleonomic processes cannot be thought of as running "only" by physical laws as well, and about what must be added to physical laws to make them applicable to life processes.

Johnston and Turvey (1980) discuss two ways of interpreting the idea of a program for adaptive behavior. The first involves a set of instructions followed by some executive component; this leads to a regression in which one needs to account for the executive function. (In treatments of ontogeny, the genetic program is treated as both instructions and executor, which hardly lessens the difficulty.) The second interpretation treats programs as implicit in the structure of the system. As these authors point out, in this second interpretation the program is an a posteriori description of the system, not an a priori prescription for it.

A last aspect of Mayr's treatment that is relevant to our discussion is his suggestion that a program may be transindividual. He considers an individual, in a "broad sense," a pair of birds during reproduction, "whose instinctive and learned actions and interactions obey, so to speak, a single program" (1976, p. 365). Locating prescriptive programs in genomes would not allow easy extension of the notion to highly organized interactive systems involving organisms of different species (symbiosis, for example, or prey-predator relationships) or even to ecosystems, which may go through fairly predictable progressions and may even, as noted earlier, restore themselves after disturbance.[4] Identifying programs with complex systems, on the other hand, is consistent with what I take to be basic about them, their various degrees of orderliness and efficacy, without postulating directive entities. Miller, Galanter, and Pribram speculate on the feasibility of extending their concept of Plan to organism-environment or multiorganism systems (1960, p. 78). They draw back a bit from formalizing "shared" or "public" plans, but give the impression of doing so not on compelling theoretical grounds but rather out of a reluctance to pursue the idea prematurely and, perhaps, out of a sort of intellectual fastidiousness.

Lorenz invokes Mayr's programs frequently in his discussion of adaptation as phylogenetically and experientially acquired information, though he sometimes seems to go beyond Mayr's concept. As examples of the "cognitive mechanism" of an open program, he cites hemoglobin increase at high altitude, increase in a dog's coat thickness during cold weather, and plant phototropism, declaring that such programs require more, not less, genetic information than closed ones (1977, pp. 64–65). He then offers another example: induction of embryonic ectoderm, which he considers an open program because the ectoderm's prospective potency (developmental potential) is greater than its prospective significance (actual developmental fate as neural tube).[5] Lorenz sees the ectoderm as containing all programs that might be used, since the organizer (the amphibian blastopore lip in Spemann's classical research) does not inform the ectoderm, but rather selects from the various possibilities in the open program. The "external impulses" that make the selection are "themselves already 'planned for,' in the sense that they are built into the programme on the basis of previous adaptive processes" (pp. 67–68). This idea that an open program contains more information than a closed one because all possibilities and their triggers must be anticipated is crucial to Lorenz's treatment of learning and other "adaptive modifications" as highly structured and directed by "innate working hypotheses." This in turn is necessitated by his insistence that learning could not be adaptive if it were not already provided for by innate mechanisms, that external influences not so constrained would harm the behavioral system much more often than they improved it (pp. 80–84).

Lorenz is correct that random external direction would rarely benefit the organism, but the reciprocal selectivity of such interactions ensures that they are never random, despite our habit of thinking of environmental influences as unordered. The structure of the reacting organism and the structure of the stimulus are both reflected in the result. (His argument parallels the frequently heard one that says well-designed organisms cannot arise in evolution as a function of "blind chance," and that some other explanation is therefore needed.) Since by Lorenz's reasoning, thalidomide administered to pregnant women at different points in gestation "selects" different types of limb formation, each distinctly nonrandom, fairly specific not only to the species but also to the developmental age of the fetus, must we say that the various defects are built into the program? Lorenz would surely argue not, since malformed or absent appendages are not adaptive. But then, what knowledge resides in the program is defined not by selectivity of response, but by what is considered normal or adaptive. To explain the normal and adaptive by that knowledge, then, is of questionable utility.

The same criticism, I think, can be made of Williams's distinction between adaptive responses and their failure (1974, chapter 3). He faults Waddington's interpretation of genetic assimilation of the bithorax character in fruit flies (originally produced by abnormal treatment and eventually, through selection for the tendency to develop the character, appearing even with normal treatment) and argues that there is a vast difference between a coordinated adaptive *response* to a stimulus and a susceptibility to display a disorganized *effect* like the abnormal body segment seen in bithorax individuals. The latter signals failure, not function, of adaptive mechanisms. In positing susceptibilities as an evolutionary source of responses, Williams claims, Waddington showed a basic misunderstanding of selection. I am certainly not denying the difference between well-formed limbs and flipperlike anomalies or between a normal segment and an abnormal one. What I am suggesting, though, is that these distinctions are made on the basis of a judgment about the phenotype itself, not about the kind of biological interactions that led to it. There is no reason to assume that such unusual "effects" can never turn out to be beneficial to the organism involved and eventually come to typify part or all of the population.

To return to Lorenz, we see that even though embryonic induction is mentioned repeatedly in his discussions of development and learning, it is not development that he is truly interested in, but adaptedness. This preoccupation, and its difference in focus from that of developmentalists, was noted by Lehrman (1970) and is important to keep in mind. When we look at these ideas from the point of view of ontogeny, they are less illuminating. How helpful is it, for instance, to say that the ectoderm "contains genetic information on all the programmes that it is potentially capable of carrying out" (1977, p. 68)? As was pointed out before, virtually every cell contains all of the genetic "information" all the time, and yet not all patches of cells can be induced in the same way, and inducibility changes with time. If, in other words, one wishes to emphasize the peculiar combination of flexibility and selectivity that characterizes such processes, can one do so more effectively by referring to the state and history of the competent tissue or to an abstract body of "information" shared by competent and noncompetent tissues alike?

Another consequence of Lorenz's depiction of embryogeny as a function of open programs is that the entire architecture of the organism then depends on the progressive integration of "environmental" information. I have no argument with this, but the implication is that species-typical morphology and physiology, traditionally considered "innate" and completely controlled by genes, are all results of *open* programs, and thus the distinction between closed and open becomes obscure, especially given Lorenz's treatment of learning as an extension of embryogeny. He distinguishes these last two principally by the permanence of embryonic determination and the mutability of learning (1977, p. 68), neither of which is a reliable generalization. Part of the importance of the induction model to Lorenz is that it demonstrates the irrelevance of the source of the "external impulse" that induces a change (p. 68). Whether the impulse originates in the organism itself or in an external stimulus, it must be foreseen by the genome if a response is to be adaptive. While the deemphasis of the difference between intra- and extraorganismic stimuli is important in that it could prepare the way for the elimination of the distinction between the innate and the acquired, Lorenz instead embarks on an infinite cognitive regress that places information on all important interactions into open programs, so that virtually every ontogenetic contingency seems to be written in the genes. We have seen that some workers faced with such contingency have tried to place *less* in the genes by speaking of hypotheses rather than instructions (while risking having to put all possible hypotheses in the genome); Lorenz places *more*. If one is unwilling to abandon the assumption that activity requires knowledge, one is caught on the horns of the ancient dilemma: How can one acquire knowledge if one can't recognize it? And if one recognizes it, doesn't one already have it?[6]

Williams also believes that more information is required for conditional reactions than for invariant ones. Part of his objection to Waddington's bithorax research was, as we have noted, that the character is a deformity, not an adaptive feature. (There is, however, reason to believe that the elaboration, and even repetition, of body segments and their parts has been an important part of the evolutionary history of many organisms. See Raff and Kaufman, 1983, chapter 8.) Even in the case of adaptive abilities, however, like callus formation, he disputes Waddington's account of genetic assimilation as evolutionary progress. (The hypothetical sequence here was from facultative callus formation under the influence of pressure and friction to obligate calluses that are discernible in the embryo.) Instead, Williams states, it is degenerative evolution, because it involves a transition from a complex conditional instruction to a simple one requiring less genetic information. "The process starts with a germ plasm that says: 'Thicken the sole if it is mechanically stimulated; do not thicken it if this stimulus is absent,' and ends with one that says: 'Thicken the soles' " (1974, p. 80). Though it is true that the facultative response must itself be explained, Williams's assumption of a correspondence between physiological and logical complexity seems unjustified. The valuable point here, I think, is the same one that Lorenz is attempting to make when he speaks of genomic plans for learning. It is that learning (or other interaction with the environment) is not random, either statistically or with respect to organismic characteristics. In the view being developed here, the stimuli and organism are usually part of a developmental system that has been recurring for many generations, and the adaptive fit between them often resembles that between parts of the organism's own body. I too am impressed by this, but believe that it is unhelpful to place more and more explanatory burden on the genes in attempting to express it.

The more information, hypotheses, potentials, and foreknowledge are placed in the genes, the more pressing is the issue of selective activation. It is here that the fundamental unity of all notions of determination, information, and control must be confronted; all these variations seem ultimately to reduce to the specification of alternatives. Whether one is speaking of determination of statistical variation or naive causation, the underlying question is always, Why this and not that? As Lorenz points out, it matters little whether the selective influence is internal or external. A sensory stimulus, after all, must be transduced into physiological events before it is biologically relevant, and internal events and materials are continuously responsive to, and exchanged with, the external world. One could argue that a basic aspect of life processes is just this: the constant but selective, progressive exchange of inner and outer.

If multiple possibilities are to be taken seriously—and they are everywhere evident in biology, from the molecular level to the ecological—the selection from the array of possibilities cannot be an exclusive function of the array itself. Nor, and this is an important point, can the array, the "potential," be strictly defined apart from the array of selecting influences. The fact that, in normal development, a multitude of selections and influences occur inside a skin does not change their mutual dependence. A structured system selects its stimulus—indeed, defines it and sometimes produces it (the state of the system determines the kind and magnitude of stimulus that will be effective, and intrasystemic interactions may trigger further change)—and the stimulus selects the outcome (the system responds in one way rather than another, depending on the impinging influence). Nativists have traditionally focused on the former, while empiricists have stressed the latter. In doing so, they have perpetuated and further polarized the opposition between fated internal structure and fortuitous outside circumstance. The mutual selectivity of stimulus and system applies to causal systems of all sorts, and illustrates the impossibility of distinguishing definitively between internal and external control, the inherent and the imposed, selection and instruction. Selectivity of response does not imply foreknowledge of the stimulus, though it does reveal something of the structure of the system.

What of the use of program analogies by other scientists? Computers do marvelous things by program, so perhaps they can explain life processes as well. Conrad (1972) acknowledges that it is possible to conceive of a self-reproducing machine, a "universal constructor which is provided with a description," and that people have attempted to apply these ideas, as well as other computer analogies, to biology. (See von Neumann, 1966. Ransom, 1981, p. III, observes that the problems of making a workable kinematic automaton were "immense" and that von Neumann soon turned to other models.) But, Conrad cautions, the resemblance between Turing machines reading tapes and genetic processes is superficial. "The DNA is not a program or sequentially accessed control over the behavior of the cell. This is because the biological process of translation does not correspond to the construction process." Though some "sequential accessing" of blocks of DNA occurs during parts of the cell cycle, this is not analogous to "sequential action of manipulable elements" in a computer program. He is doubtful that any model in which units react only to the specific output of other units, rather than to "global properties" of groups of units, can be adequate to biological processes. He goes on to discuss in more detail the differences between machine information processing and life processes, adding warnings about oversimplified notions of simulation. With refreshing straightforwardness he notes that DNA describes the primary structure of a molecule, not the behavior of a cell (Conrad, 1972, pp. 223–226), something that is too easy to forget in the midst of heady talk of physiological algorithms.

In the context of a sophisticated discussion of information in biology, Gatlin pursues the computer analogy at some length, showing less skepticism about its utility than Conrad. Still, she concludes that the genetic code is but "a small subroutine of a master program which directs the machinery of life" and that she doesn't know the language of the master program (1972, p. 118). She might also wonder where it is to be found; could such a notion have any meaning unless it were identified with the totality of structure and function of the organism and its surround, and thus were itself subject to ontogenetic change?

Even without speculating on master programs, it should be evident that gene transcription and translation in no way represent instructions for building a functioning body, though they are surely part of that process. Whatever determines when a given gene will be transcribed, it is something other than that gene itself.

Goodwin (1970, pp. 3–5) asserts that DNA doesn't function as a computer program because a cell about to change its state does not compute that state, refer to the genes for conditional instructions, and then change state accordingly. "If one is going to use this language, it is necessary to point out that the relevant algorithm for the decision, while ultimately specified in part by the DNA, actually operates at the moment of decision through enzymes, other macromolecules, and metabolites." Again we see that the description becomes accurate just at the point at which it departs most from the original image. The "program" fits biological processes only when it is no longer a set of instructions directing a process, but is identified with the process. I recently heard a molecular biologist talk about the developmental totipotency retained by many differentiated plant cells. Such cells will, under certain conditions, produce a new plant. They retain, he said, the program for producing the whole plant. What I found striking about this offhand remark was the separation of the notion of "program" from the genome proper. It is possible, that is, to retain the original gene complement without retaining full developmental competence.

Ransom, in a book on computer simulation of development, begins with a standard presentation of the genetic blueprint as reconciler of preformationist and epigenetic approaches, but continues that the information must be "unlocked in the correct sequence" and that the blueprint must "interact with the environment," which will then affect the direction of development (1981, p. 2). As the description becomes more elaborate, it comes rather to resemble the exquisitely sensitive and skillful improvisation of a house without blueprints. To carry the fantasy further, it would have to be a house with someone living in it from the very beginning, constructing itself on a continuously changing site with a shifting pool of materials and tools. It would be responsive to some of these changes and unresponsive to others. Ultimately the metaphor is stretched to the point of surrealism and the imagination fails. One wonders, incidentally, whether it is accidental that an author so well versed in computers and programs chose the blueprint over the genetic program.

Though the genome as computer program shares many of the problems encountered with other cognitive metaphors, it presents some special ones as well. Having entered the intellectual public domain, it seldom functions to point out a limited similarity between two phenomena, but instead is taken as a statement of fact. (Hence the pickiness of some of the above criticism; if you don't think the genes are really a blueprint it is not necessary to point out that they are not blue, but if you think they really are a program, closer examination is perhaps justified.) Computers and their programs are also, by and large, the results of attempts to simulate and extend human intellectual processes and other complex phenomena. Their features develop in a complicated dialectic between technology and the fields of study in which they are used. Precisely what a program is like and what it "does" is therefore changing every day. Burks (1966, pp. 7–8) mentions the time-consuming process of manually programming ENIAC, the 1940s high-speed electronic digital computer, and describes von Neumann's improvements of the procedure by allowing programming without setting switches and connecting controls with cables. Miller, Galanter, and Pribram (1960, pp. 41–42) observe, "It is amusing that so many psychologists who abhor subjectivism and anthropomorphism unhesitatingly put telephone switchboards inside our heads," and point out that early switchboards required humans to make the connections. Later, automation mimicked those humans' actions. (Similarly, writers who would blanch at placing a technician in our bodies to build them, wire them, and make them run, blandly put programs in the technician's place.) The same authors (Miller, Galanter, and Pribram, 1960, p. 16) identify their concept of Plan with computer programs only hypothetically, suggesting that a program be thought of as a "theory about the organismic Plan that generated the behavior" being simulated.

Programming innovations described by Ransom (1981, p. 108) allow modeling of cell reproduction and differentiation. These features presumably improve the ability of the researcher to do justice to biological processes. I do not doubt that, with time, theoretical and empirical advances in psychology, biology, and other fields will be accompanied by, and perhaps aided by, increasingly effective computer simulations. What is important to remember is that a program is itself a sort of model or metaphor. It has features like look-up tables and control lists for "deciding" outcomes *just because a computer lacks the biological structure and dynamics of an organism.* Hofstadter observes that the many rules and strict formalisms employed in computer programming are ways of telling an inflexible machine how to be flexible (1980, p. 26). Biological processes, however, frequently *are* flexible (by nature, if you will), and are the very phenomena the programmers are attempting to imitate. A program for simulating ontogeny, even a small part of it, would have to include not only genomic structure but also descriptions of all the conditions, parameters, and interactions, internal and external to the organism, that constitute the developmental system in transition. In the biological system, the "decision" or "rule" needn't be programmed symbolically; "rules and decisions" are simply our anthropomorphic descriptions of the events we observe. The events themselves are "controlled" by reciprocal selectivity at various levels of analysis, codetermined by the entity and conditions, state, and input, whether or not they involve a change in the pattern of gene transcription in cells.

If processes are "programmed" when they can be analyzed and simulated in this way, then all biological events are programmed, and the concept of program cannot be used to distinguish the innate from the acquired, the adaptive from the nonadaptive, the inevitable from the mutable.

In introducing the basic vocabulary of the computer modeler, Ransom defines "rules" as interactions between components (1981, pp. 9–11). They are, simply, kinds of events. Rules and instructions, in other words, are what we formulate on the basis of observation of the universe to be simulated. They are what we must use to produce results that resemble the operation of a system that runs without "rules" as we know them, but rather produces orderly outcomes by virtue of its evolving nature and its interactions. The regularity we describe, because it is always multiply determined and is a function of the history of the system, cannot reside in a component of the system. It is the *result* of the operation of the system, not its cause (except insofar as results become causes by altering the system itself).

There is a subtle, repeated process at work here. Order in a process is perceived and formulated as descriptive rules. From these, prescriptive rules are derived and imposed on a mechanical medium to allow simulation of the original process. The prescriptive rules are then projected back into the original process as cognitive agents, programs, accounting for the original order in terms of the simulated order. The working of the original is then said to be "like" that of the imitation, and therefore due to the same kind of intentional control that created that imitation. To say it another way, order is abstracted from one system and imposed on a second, then the imposed order-as-program is abstracted from the second and projected into the first. Rather than assuming that ontogenetic processes fit our notion of programs, we should be asking (and in fact people involved in computer simulation *do* ask) whether our notions of programs do justice to ontogenetic processes. We do well to remember that the word "model" is ambiguous, referring at times to the original, as in artist's model, and sometimes to the replica, as in model airplane. In a letter written in 1649 to Henry More, Descartes says, "since art is the imitator of nature, and since man is capable of fabricating various automata in which there is motion without cogitation, it seems reasonable that nature should produce her own automata, far more perfect in their workmanship, to wit all the brutes" (cited by Grmek, 1972, p. 186).

As we contemplate the nature in and around us, the argument from design is ever present. When we remember that our cognitive metaphors are motivated by it, as is the case when we say that an embryo develops as though it had a full set of instructions, all is well. When we forget, we entrap ourselves in the worst kind of pseudoexplanation. What is "worst" about such explanation is not that it explains nothing, but that it seems to explain everything. When preexisting plan is used to explain what is, what *is* becomes *necessary* (or at least natural, normal, or hard to change). As we will see later, in precisely those situations where we need to be most circumspect about definitions of the normal, desirable, and necessary, explanation by programming narrows and defines the field for us, prejudging what we should be mustering our full moral and intellectual powers to decide or find out for ourselves. The subtlety of our deliberations is not apt to be increased by our readiness to translate "it's genetic" or "it's biological" as "it is written."

We return from this excursion into the world of automation with essentially the same lessons we have gleaned from our other investigations. A cognitive metaphor that is meant to stand for the whole organism, with its species-typical processes and outcomes, may be a useful form of shorthand. An attempt to explain development by such means is unintelligible at best, and theoretically, politically, and morally dangerous at worst because it encourages us to predict limits in situations whose critical parameters may be unknown to us. If, on the other hand, causation, direction, and control are seen as interacting in ongoing, integrated processes, fidelity to the metaphor is sacrificed as intelligibility is regained. A blueprint, plan, or program that truly reflects ontogenesis, that is, is no longer a blueprint or plan, but simply a description of phenotypic state at a given time, competent to respond in certain ways, shaped in certain ways by its history, and surrounded by an organized milieu. Some authors do occasionally place causal agency in phenotypic processes, usually when they are not generalizing but are describing a process in some detail or following the implications of an image with particular care. This seems to happen almost inadvertently, however, as the general formulation always names the genome, not the entire developmental system, as the fundamental source of order.

E. Information: Communication, Controls, and Constraints

In addition to being apparently more innocuous as a metaphor than some other variants, "genetic information" has a powerful cachet acquired by its association with physics, information theory, and computer science. It is very much in vogue among life scientists. In a striking reversal of metaphoric polarity, Campbell, in discussing the historical changes in the meaning of "information," claims the term took on active connotations in the 1940s, when information came to be seen as influencing the material world "much as the messages of the gene instruct the machinery of the cell to build an organism." The active gene is now the prototype that informs our understanding of information. He continues, "Nature must be interpreted as matter, energy and information" (1982, pp. 15–16; see Jacob, 1973, p. 95, for an almost identical statement). Eigen and DeMaeyer speak of life processes as being "intelligent" at the molecular level, and as comprising "the storage, readout and transfer of information, the processing of information in logical operations implemented in molecular interactions (1966, p. 245). Gatlin (1972, p. 1) goes so far as to state that "information processing system" constitutes an operational definition of life, and in cognitive psychology the nervous system as information processor is standard.

In physics, information, as thermodynamic order or available energy, is contrasted with entropy. Concern that biological processes in some way violate the Second Law by reducing entropy has been largely laid to rest, but the increase in complexity effected by such processes continues to puzzle and provoke. Wicken (1979) discusses the compatibility of "entropic" (thermal order decreasing) with "biologic" (organizing) processes, arguing that increase in various levels of organization can be seen as "thermodynamically favored" ways of reducing statistical order. (Except where thermodynamic order is under discussion, however, I will use "order" and "organization" interchangeably.) He points out that unlike a closed physical system, which requires insulation from the outside world, open biological systems require "kinetic stability," which is maintained through "transformational interaction" with the environment. It is these kinetic considerations that are of primary importance in natural selection for functional organization (pp. 294–297; see also Prigogine and Stengers, 1984, for a treatment of emergent structure; and Schrödinger, 1967).

I. INFORMATION AS COMMUNICATION

With the elaboration of information theory in communications engineering, information was detached from exclusively thermodynamic considerations and employed in statistical treatments of messages moving through channels from sources to receivers. Campbell reports that Claude Shannon, one of the key figures in these developments, was uneasy with the popularization and overapplication of information theory, a concern that Campbell himself, in his enthusiasm for the potential of the information concept as a "universal principle at work in the world," appears not to share (1982, pp. 16–17). In what is probably the inevitable outcome of our passion for the grand unifying vision, Foster describes the universe as one vast intelligent computer, and everything in it as data. Data, in fact, replace energy as the "prime universal commodity" (1975, p. 441).

What is fundamental to the communications concept of information is its purely statistical nature; content and meaning are not considerations, but predictability is. The amount of information transmitted, that is, depends on the number of alternatives; an event that reduces no uncertainty carries no information. In much the same way that a variable becomes a determinant when it accounts for a difference in outcome, a message conveys information when it distinguishes one possibility from another. In both cases, quantification is possible only in the context of a well-defined system, and that which doesn't vary can neither account for variability nor convey information. That which is held constant, however, like controlled variables in an experiment or redundancy-increasing constraints or rules in a code, constrains by limiting possible variation. However much biologists may invoke entropy, negentropy, and Maxwell's demons,[7] this content-free, probability-based concept of information rarely plays a part in their discussions (though the way information is actually used carries hidden statistical implications, as will be seen later). If it did, "information" would be used in very different ways and with much more circumspection (Gatlin, 1972, p. 19; Hinde, 1968, 1970, p. 428; Klopfer, 1969; Waddington, 1975, p. 215).

If the basic components of the communication model are examined, it becomes clear that communication theory is generally invoked in an extremely loose manner. What corresponds to the source, for instance, that selects symbols from a set and encodes the message? One might argue that the zygote represents a source whose DNA is assembled at fertilization, or that the process of reproduction, or the species, or the gene pool is the source. The channel might be ontogeny itself. Where communication theorists deal with relatively closed systems, however, in which information may be degraded but not increased, ontogeny is an open one in which energy is dissipated but complexity often increases. The engineer's message can only be harmed or lost by transmission, while the ontogenetic "message" is created in transmission.

Nowhere, furthermore, have I seen mention of the receiver, even though its characteristics are an important determinant of the amount of information that can be said to be transmitted. Wiener emphasizes the difference between information per se and information that enables an organism or machine to act effectively. The latter, "semantically significant information," is that which reaches an "activating mechanism" in the receiving system; it is determined by the structure of the receiver (1967, pp. 127–128). The significance of any information influencing developmental processes, in other words, whether from genetic transcription or from other events, is defined by the developing system itself. The phenotype as receiver is implied in many treatments, though occasionally "the environment" or "natural selection" is conceived as a kind of interlocutor in a dialogue of adaptation and survival.[8] The message, on the other hand, has been variously characterized as proteins (Campbell, 1982, pp. 92–93), a collection of enzymes, structural plans for a cell, and organism (Jacob, 1973, chapter 5).

While it is plausible to apply communication concepts to the reproduction of nucleic acids and the manufacture of proteins, this would limit concern to the reliability of transcription and translation, and perhaps to the stability of the genome. Working along these lines, Gatlin (1972) presents an elaborate analysis of redundancy in DNA. On the other hand, extension of the analogy to encompass heredity in general, which Ransom considers to be "far-fetched" and fraught with difficulties introduced by nongenetic influences and changes in the genetic material (1981, p. 117), begs the question of exactly what is passed on by heredity while seeming to answer it. In addition, though most people seem to agree that organic "information" (organization or complexity) increases with development, no one knows quite how to describe this, much less how to quantify it. In information theory, quantification requires knowledge of the number of possible messages, a datum not available to us.

Having identified the message variously with chemicals, rules, instructions, and organisms, Jacob states that heredity is the "transfer of a message repeatedly from one generation to the next. The programmes of the structures to be produced are recorded in the nucleus of the egg." He quotes Schrödinger on the genetic code as both plan and executor (1973, pp. 251–254).[9] A cell is not the same as a group of chemicals, however, and neither is an organism. The "genetic message," as Jacob himself admits (presumably when construing message as genes), can do nothing without a functioning cell, and even within a cell, it "plays the passive role of a matrix" (p. 278; recent research gives a less passive impression of the genome, but Jacob's point stands). It is indeed an entire cell that is reproduced, not because the genes contain instructions for building it, but because any inheritance involves passing on DNA and all the cellular and extracellular structures, processes, and materials necessary for its exploitation.

2. INFORMATION AS CAUSAL CONTROL:
VARIATIONS AND CONSTRAINTS

In such information-based treatments of biological processes, then, a transition is quickly made from formal considerations of numbers of alternatives and reduction of uncertainty to the familiar concerns with causal control of development and adaptation. A related transition is that between the restricted analytical use of genetic determination or control as explanation of *differences* within or across populations (a use that, being based on correlation, is compatible with information theory) and as explanation of *kinds of development* (maturational, canalized, governed by epigenetic rules) and *characters* (innate, biological, programmed, phylogenetically derived). This shift from information as a *measure of unpredictability* to information as an *explanation of predictability* is frequent and unreflective, and, given the technical legitimacy of terms like "determinant" and "control" in certain contexts, lends a similar legitimacy to their misuse. If genetic information is used to convey the impression of intelligence and foresight an observer receives from adapted structures and functions, the comments made earlier on images and knowledge in the genes are pertinent. Since adaptive traits (or, for that matter, nonadaptive ones) cannot be analyzed into genetic and nongenetic parts, such notions of information make little literal sense, but they may communicate the beauty and effectiveness of a creature's design.

The connection between information and causal control, while implicit in such adaptedness-as-knowledge metaphors and while intimately related to the notions of plan and instruction, merits specific attention because genetic plans may at least be recognized as figures of speech, while "genetic basis" and "control" seem instead to be straightforward technical terms. It makes sense to ask, then, just what they could mean.

Jacob, moving from a brief treatment of information in physics to cybernetics and language, characterizes information as "the power to direct what is done" and continues, "Interaction between the members of an organization can accordingly be considered as a problem in communication" (1973, p. 251). According to Miller, communication "occurs when events in one place or at one time are closely related to events in another place or at another time (1973, p. 3); communication and causation have become difficult to distinguish.[10] The same author, with Galanter and Pribram (1960, pp. 27–29), describes the TOTE (test-operate-test-exit sequence), the feedback loop that forms the basis of their concept of the Plan. They point out that the loop may be concretely considered as a flow of energy. At a more abstract level, they say, one can think of information as flowing through the loop, and the Shannon-Weaver notion of transmitted information is described as a correlation between input and output. At a yet more abstract level, they continue, it is *control* that moves through the loop, and they offer as an alternative description of this process simply "succession" — that is, a sequence of events. They refer to the passage of control in a computer from instruction to instruction and liken it to certain human activities in which an orderly progression occurs among controlling stimuli. Control here is mobile and contingent, dependent on the outcome of the "test" and "operate" components of the loop. Wiener notes that information is "more a matter of process than of storage" (1967, p. 166; see also Hailman, 1982, on ontogeny as changes in control, and note 1 to this chapter). This is all consistent with what was said earlier about programs. One may seek to simulate a complete process, but control cannot then be identified with only one of the aspects of that process. (It is possible to reverse the metaphor, taking the cell as programmed computer and the genome as input; in any case output is jointly controlled. See Hofstadter, 1980, p. 547, on the absence of clear distinction between program and data, interpreters and processors.)

The more one compares cells and other biological entities to computers, the clearer it becomes that it is not such contingent, causally interdependent processes that most people want to evoke when they say a character is genetically controlled, but rather a sense of reliability, even inevitability and immutability—the gene as source, not of variation, but of null hypotheses about "environmental influences" and "malleability." As long as both input and program are unchanged, of course, a properly functioning computer will produce the same results again and again. If the results vary, the variation is "under the control" of whatever varied in the process. The *sameness* of repeated outputs, however, is not attributable in such a simple way. It may require precisely the same program and data or it may not. It certainly does not indicate what may or may not be an effective variable in other circumstances. This is the old problem of the uninformativeness of negative research results. Presumably the lack of functional relations is due to "constraints," but one doesn't know what they are or how they work. The influence of a constraint can depend on other factors as much as the influence of a variable can.

Control here is roughly equivalent to naive cause, the difference in a factor that makes a difference in another, the variable with which an effect co-varies. Constraints may often be thought of as variables that might have varied but didn't, and whose present values function to limit the scope of existing variation. In addition to being a source of variation, then, genotype is frequently seen as setting limits on phenotypic variation, and these constraints are what certain writers seek to place in the genome, in the form of ranges of possibilities, programmed potential, and the like. But such constraints are aspects of developmental systems. While it is true that a gene that does not exist cannot be transcribed, an existing one might not be transcribed, either, because of genomic structure, the history of the cell, or current conditions; the implications of this lack for the organism will naturally vary with the rest of the genotype and with developmental circumstances. What is an *effective* genetic constraint depends on the rest of the developmental system as much as the effective stimulus or effective "information" does. Environmental constraints are defined in the same manner; it is the organism that synthesizes all these constants and variations, some of which leave their traces as internal constraints — the structure and function of the phenotype.

An intimate relation exists among the ideas of constraint, redundancy, rule, complexity, and structure. (John von Neumann, 1966, p. 60, observes that redundancy makes complexity possible.) They all imply a restriction of possibilities, the channeling of activity, the interaction of causes, and are thus fundamental to vital phenomena. Constraints are as interdependent as sources of variation are. If genetic information exerts control by variation (accounting for variance) and by constraint (limiting variance), then environmental information, both variable and stable, must exert control in both ways, too. But then phenotypic constancy, through an individual's life, or within or across populations, cannot be attributed to information from one source or another (cf. Chapter 2, discussion on persons and situations), and certainly cannot justify the prediction of constancy under other conditions. The developmental significance of any and all of these roles is a function of the total ensemble of variables, conditions, and interactions in question. All the caveats that attend the use of concepts referring to the statistical determination of variation, then, must also be in force when information as variation or as constraint is invoked, however generally, casually, or metaphorically. By considering "genetic" to be anything that can vary with genotype when genotype is varied (genetic determination of variance) *and* anything that remains constant if genotype is not varied (genetic constraint), of course, one covers virtually the entire field.[11] By switching between these two senses without explanation, many treatments of the issue are hopelessly confused in nonevident ways. Add to this the tendency (discussed in Chapter 4) to explain differentiation of cells with the same genome (*non*genetic determination of variance) by that shared genome, and the stoutest heart falters.

The rigor of a strict communication theory treatment of information has been abandoned, apparently for a nonstatistical, causal explanation of development. A second look, however, reveals that the causal explanation ultimately reduces to an implicitly statistical one (statistical in the sense that variation and lack of variation are attributed to factors that did or did not vary and predictions are made on the basis of these attributions), but freed, by metaphorical license, from the usual restraints on technical usage.

F. Code

Though the genetic code has become part of everyday language, a brief discussion of this variant is necessary. It is often combined with others, as in coded information or instructions. It is clear that whatever problems arise with these other versions are not solved by claiming that they are encoded in the genes rather than literally being present in them. A code does not alter the content of a message, only its form.

Gatlin (1972, pp. 125–129) describes a code as a way of mapping a domain onto a counterdomain. If several elements in the domain code for (map onto) the same element in the counterdomain, a degenerate relation exists. This is the case in the genetic system, where several codons can code for the same amino acid. Ambiguity, on the other hand, exists when an element in a domain maps onto more than one element in the counterdomain. By Gatlin's definitions, a relation that is both degenerate and ambiguous does not even seem to be a code.

This formal requirement of a code should be kept in mind. When Lumsden and Wilson, for example, speak of "compressible systems" that can be "described or regenerated by a set of rules or instructions much shorter than the shortest direct description of the system itself," and go on to say that it is possible "to recover the original system in detail" (1981, p. 345), they seem to be speaking of some sort of mapping. But though they suggest that human social phenomena fit this model, their unilinear causal scheme (genes to epigenesis to individual behavior to culture [p. 343]) is not consistent with the degeneracy and ambiguity found in ontogenetic processes at many levels. In the phenocopy, several genotypes are associated with the same phenotype (Oyama, 1981); this also occurs whenever there is genetic variability that is not reflected in phenotypic variability — that is to say, very frequently. Mapping ambiguity is observed whenever a genotype is associated with several phenotypes; this would include the differentiation of tissues during ontogenesis and the norm of reaction for a given genotype. The same argument can be made for any particular environmental variable or event. Many stimuli may be functionally equivalent for an organism, and a given stimulus may call out a variety of responses (or none) from an organism at different times or from different organisms. In ontogenetic processes, then, neither developmental interactants nor phenotypes are uniquely recoverable from each other. Where, then, is the code?

G. Metaphors and Development

That conceptions of causality and knowledge underlie this heterogeneous collection of analogies is not surprising, though the persistence with which new versions are generated as old ones are discarded deserves some attention. There is, after all, nothing disreputable about being preoccupied with causes and adaptedness, and most of the workers quoted in these pages have made important contributions to understanding both. What seems to drive people to embed their discoveries in sequences of similes is a diffuse sense that their analyses are incomplete, that general synthesis is required to unify the congeries of little miracles. But final causes are not permitted. Hence the growing pantheon of miniature gods in mufti, as the cautious avoid discredited ways of speaking but continue to seek the means to express their intuitions about life processes — that they are uncanny in their intricacy and efficiency, and that they reek of intelligence.

Surely repetitious at times, this examination of variations on the theme of the homunculoid gene was compiled because, though most have been criticized at one time or another, they have never, to my knowledge, been treated together. The problem with piecemeal criticism is that it tends to be ignored, or, if not, to be assimilated piecemeal. A writer with new scruples about "innate information," for instance, will transfer all or some of its meanings to a genetic program; or one who is disillusioned with coded instructions will turn to coded hypotheses; and few have any qualms about distinguishing traits or abilities that are biological from those that are not. The game of terminological chairs will continue as long as we look for places to set our assumptions about sources of order, higher agencies that organize all the analytic causes and effects. This is why it has seemed to me imperative to follow the course of these images with care.

There is a certain poignancy to our dilemma. We turn to these cognitive-causal metaphors in our zeal to avoid supernatural explanations, but because our ways of thinking are too conservative to match our antimetaphysical resolve, we find that the form and content of our statements have often not changed with our vocabulary.

In addition to our intuitions about development and adaptedness, however, we reveal other preoccupations when we doggedly return to these questions of primary causation and fundamental nature. These preoccupations, which we will examine in the next chapter, are ultimately more moral and philosophical than empirical, having to do with freedom, necessity, the natural, and the good. Not knowing whether this is the only possible world, we avidly look at the way its creatures, including ourselves, got this way, hoping to learn what might be and what should be.

After considering these concerns in more detail, I will attempt to assemble an alternative way of thinking about ontogenesis and phylogenesis, a way that allows us to think about genes and environments without elevating them to principles and pitting them against each other, and thus losing sight of the very organisms that were the reason for our inquiries in the first place.

6 The Ghosts in

the Ghost-in-the-Machine Machine

I am aware that I have occasionally invoked Gilbert Ryle's ghost in the machine without devoting much effort to making my philosophical intentions clear. Ryle was concerned, in *The Concept of Mind,* not with the role of genes in development, or even, really, with the nature-nurture problem, but with the mind-matter dichotomy and the difficulties he felt it posed for an adequate treatment of behavior and experience. His "ghost in the machine" stood for the causal entities traditionally used to explain will, perception, knowledge, and other mental functions, and he admitted that he used the phrase with "deliberate abusiveness" (1949, pp. 15–16).[1] Since my interest is rather in ways of thinking about development and about the meaning of the "biological" in our lives and studies, it is appropriate at this point to explain, without venturing into excessively deep philosophical waters, my adoption of a phrase that may seem ill-suited to the task at hand.

A. Perennial Pairings

When we become aware of the longevity and the protean persistence of the nature-nurture opposition, in spite of constant efforts to resolve or redefine it, in spite of the incompatibility and even mutual contradiction of its parts, in spite of the conceptual disorder on which it rests, our puzzlement resembles that of the observer of biological processes. The raw materials seem inadequate to explain the phenomenon in all its complexity and vigor. Last rites are pronounced over the remains of the dispute with tedious frequency by authors who resurrect it in the next paragraph, clothing it in new terminological raiment and surrounding it with ever more refined theoretical arguments. Everywhere one looks, approaches continue to range themselves in pairs. Neuronal networks are hard- or softwired (Jacobson, 1974, 1978, on Class I and II neurons), views of behavioral development are biased toward control mechanisms or interaction (P. P. G. Bateson, 1976), and in immunology, selective models have triumphed over instructive ones (Jerne, 1967; see also Jacob, 1982, pp. 13–18, 38–39, on other instances of the selection versus instruction opposition).

We have already noted that the controversy over embryonic preformation and epigenesis became progressively more subtle, until it was difficult to say precisely what separated the two views; nevertheless, the dichotomy outlasted the particular arguments. The examples could continue indefinitely. The point is not that these distinctions are intellectually useless. Often they motivate valuable discoveries, though they may also encourage sterile research or misleading interpretations. It is rather that they are so similar, even while the phenomena to which they refer are so diverse, and that, explicitly or implicitly, they have come to be tied together by the image of the DNA fiber and a questionable set of assumptions about the nature of biological processes and products.

It is tempting, when faced with such a complex and yet reliable phenomenon, to find a single reason for it, even to posit a program for it: perhaps a gene "for" anthropomorphic attribution or, after Lumsden and Wilson, "genetically underwritten epigenetic rules" for reification (1981, pp. 344–345), or yet again, an "innate tendency to range the elements of the phenomenal world in pairs of antitheses," an "*a priori* compulsion to think in opposites," particularly marked, Lorenz claims, among the Germans (1977, p. 179). Needless to say, such shorthand locutions for "these tendencies seem to be characteristic of the species" are already an extrapolation from phenomena that may be more culturally and historically limited than we think, and they are uninformative about the how and why of this fact of our intellectual history. Even if the tendencies should prove to be as common as these authors claim (perhaps not entirely seriously), in order to be relevant to the issue at hand they would have to be a good deal more specific; one suspects that even the most enthusiastic devotee of genetic programs would quail at one for nature-nurture disputes. Very general intellectual processes may in fact underlie this kind of opposition, but though I will suggest some candidates of my own, they do not in themselves constitute a sufficient explanation, and eventually themselves must be described and explained.

Some of this persistence of old oppositions is surely due to the history and sociology of modern science, as investigators train and argue with each other and attempt to describe the wider theoretical import of their findings in a way that integrates them into existing frameworks, and, increasingly, must justify the funds and institutional support devoted to their work. More generally, arguments and data must be couched in a common vocabulary to allow comprehension, and there is frequently great pressure to claim wide scientific and/or practical relevance for one's work. The concerns and political struggles of the larger society thus come into play in several ways. Scientists not only reflect current ideologies and social relations as children of their times. They must also articulate their theory and practice with those ideologies in more or less explicit ways when called upon to explain themselves. They are often in great demand, in fact, to interpret reality and make policy in ways that are congenial to the contending powers. It is hardly accidental that the successive waves of preoccupation with biological and environmental determinism have coincided with periods of turmoil over race, class, and gender.

In polemical exchanges, distinctions become finer and finer as each side uses the other's examples and formulations for its own rhetorical purposes; thus doctrine, taboo, and verbal talisman are born. Meanwhile the conceptual center sometimes shifts so that quite different arguments are taking place under the same banners. The difficulty of defining political conservatism and liberalism with any precision, and the paradoxical vigor and resilience of the labels, presents an analogous picture. Issues and ideals, as important for their symbolic significance as for their practical implications, mingle, merge, and coexist in mutually contradictory alliances, but the parties persist. Ransom, in fact, describes opposing camps in developmental biology as a right wing of embryologists who place their faith in small numbers of "wonder chemicals" to explain differentiation and a left wing of workers devoted to complex interaction (1981, pp. 188–189).[2] At some points the appearance of unity in the ranks is bought by glossing over considerable differences, while at others, particularly at the boundaries of competing groups, subtle differences are magnified. In science, as in politics, major conflicts frequently turn on alternative interpretations of the "same" facts. This is why much of the preceding discussion has seemed so abstract; it is as much about how we *think* about what happens as it is about what *happens*. The way we think about the present, however, influences the way we conceive of the future, and thus, to some extent, the future we are likely to create.

At the same time that important differences and inconsistencies among the opposing positions are obscured by categorizing them as versions of the nature-nurture dispute, a more diffuse set of concerns binds them together. What seem to be questions about genes and environment are generally resolvable to distinctions between entities that appear to do things (or events that happen by themselves) and entities that have things done to them (or events that are made to happen): roughly, subjects and objects. Hence the complex relations between this issue and those of freedom and control, reasons and causes. We view the genetic as planned and necessary and thus search the biological for clues about our own nature and fate. Before we turn to these latter concerns, however, a closer, though still informal, look at the subject-object relation is in order.

B. Subjects and Objects

What largely defines our experience of ourselves and the world is that we are subjects. In our society, growing up often means putting up with being treated as objects and enduring our lack of freedom by promising ourselves that someday we, like grownups, will do only what *we* want to. Dependence and passivity are viewed as infantile characteristics, inappropriate for the model adult (at least the adult male). Surely it is not accidental that we have tended to describe evolution as a progression toward a species that is unique in its independence of and ability to control the environment. This is a conception that both exaggerates our role as detached subjects and denies our objectlike aspects. (Merchant, in a 1982 paper, discusses the implications of the subject-object split for our science.) We perceive, we know, we act, and the distinction between subject and object seems crucial to these experiences. We perceive objects, including ourselves, knowledge appears to require a distinction between knower and known, and action implies agent and, in the archaic-sounding vocabulary of linguistics and philosophy, patient. We have reasons for acting and cause our own actions, effecting change in the world. These ways of organizing reality are so fundamental to us, at least in the Western tradition, that when we attempt to come to grips with our own mental processes, we do so by placing another "we" inside us, the mentalistic ghost in the machine. When, furthermore, we ask about the origins and character of this ghost, we immediately confront both the determinism—free will issue and the nature-nurture opposition. Is the mind effect or cause, is it formed by our experiences or does it form them? Into this set of familiar, interlocked puzzles intrudes the gene. It has a stable physical configuration (though recent advances in molecular biology are changing this assumption), fits existing ways of theorizing about biological constancy and change, and shows continuity, albeit complex, between generations. It is an instantaneous candidate for a unitary explanation of the recurrent miracles of ontogeny and adaptedness: the inner creator subject, immortal, immutable, and insulated from the untidy world outside.

When we contemplate development, what we observe on every side is evidence of perception, knowledge, and volition. The sensitivity of developmental processes, now often expressed in cybernetic terms, is seen whenever adjustments are made for variation in materials or conditions. Adaptedness of the organism to its surroundings is taken as an indication of stored information, and the events of ontogeny imply intentional constructive activity. The reliability of biological processes seems finally to demand a planning intellect.[3] An input of oxygen and milk (or later of oxygen, beer, and pretzels) seems not to explain the constitution of a human being, so these are conceptualized as passive raw materials to be used by a constructive entity responsible for the reliable creation of a particular kind of order from apparent disorder. And how to account for the subject-like characteristics of such an entity? In the same way we account for the creation of the macrocosm of which we are microcosm, or for the origin of our own actions: by conjuring up a being basically like ourselves, but with critical qualifications. God is like us but more powerful and kinder, the man or woman in our heads is like us but invisible, while the gene is like us but smaller and smarter. Supersubjects with their object aspect deemphasized, they create order from material or sensory chaos, structuring the world and manifesting prodigious foresight in arranging harmony among the parts. While, as was noted earlier, the genes (or natural selection, which is the answer to the inevitable question, Who made the genes?)

are invested with Godlike qualities, ultimately it may be that God and the soul, as well as the genetic architect in the cell—the Holy Ghost, the ghost in the machine, and the ghost in the ghost in the machine, are all our own spectral likenesses, projected outward and inward to aid understanding. We explain order anthropomorphically. When we create order, we draw on our knowledge according to certain designs and reasons, and we operate on the world. We do not always fulfill our intentions, but there is rarely any doubt in our minds about the difference between what is an intended result and what is an error or an accident.

A set of related mindlike properties, then—temporal continuity, causal potency, ability to impart order according to prior knowledge and plan —are used to explain the development and design of organisms. (See also Burian, 1978; Klopfer, 1969; Tobach, 1972; Weiss, 1966, 1967, 1969, 1978.) In addition to the appeal of anthropomorphic explanation (simplicity, vividness, emotional resonance), there are historical and philosophical bases for our creation of the gene in the image of God. The notion of evolution was quickly perceived as a potential substitute for divine creation, though not everyone agreed (or does today) on the mutual exclusivity of the two. Philosophers of biology struggle to recast the ideas of purpose, design, and goal in nonmetaphysical, nonvitalistic terms. It is here, incidentally, that machine analogies are so attractive. They seem to promise immunity from charges of mysticism. As Midgley notes, however, programs are quite "equivocal" with respect to an implied designer. She attributes the prevalence of programming metaphors to the usefulness of computers as tools, to their status as "popular cult objects," and to the "superstitions" that surround their deployment (1980, pp. 72–73, 282). Midgley also suggests that the symbolically important up-down (God-earth) dimension that was threatened by Copernican heliocentrism was soon translated into a future-past one by the concept of evolution. Natural selection was relied upon to propel us from the contamination of our lowly animal past toward a glorious, improved future purged of animal weakness, evil, and passion (pp. 197–198). The Great Chain of Being accomplished a seamless segue to the Evolutionary Ladder.

A whole family of related philosophical dualities, then, prepared us to handle the qualities of vital phenomena in these ways. Mind, reason, form, necessity, essence, and universality, as opposed to body, passion, matter, contingency, appearance, and individuality, constituted not only a ready and convenient conceptual framework for the natural sciences, but a virtually inescapable one. Though these dualities have different histories, arguments, and spheres of relevance, and though treating them as though they were merely facets of a single question does violence to them all, I would argue that just such a disrespectful conglomeration has occurred in the nature-nurture opposition. I would further argue that this opposition does as much violence to the numerous biological problems grouped under its aegis as it does to its philosophical underpinnings.

But look at the benefits. The gene as creative mind made intelligible what was most disquieting about living things while apparently absolving us of the charge of vitalism. As human intention explains orderly, sequential, goal-directed action, action that can be flexibly altered according to circumstances and opportunity, as choices made with foresight and purpose explain constancy, change, and variation in human conduct and creation, so does the gene explain constancy and flexibility in biological form and process. When I carry out a plan, adjusting my activity to obstacles, that aspect of the accomplished action that reflects my adjustment is not a reflection of my plan in the same way a perfect outcome is, though both are consequences of my intelligent activity. Even the absence of perfect outcomes does no harm to our belief that a plan underlies the variable results. When we survey the living world, our sense of order is so strong, both because we are so adept at perceiving patterns and because we have a philosophical and historical tradition of considering abstracted form more fundamental and real than actual form, that what variation we do see does not threaten the assumption of a plan any more than my detour around a pothole on the way to the store makes you wonder whether my movements were intentional. Indeed, if I did not skirt it, you might conclude that I was not using my head. Why God would permit error and deformity in the world is for some a painful and difficult question. Since we do not expect of the gene the benevolence we expect of God, such things are dismissed as mutations, random environmental effects, or acquired "phenotypic adaptations." While the exact etiology of the defect or variation may be obscure, what is not in doubt is that it is neither the result of the species plan nor a threat to the idea of such a plan.

One of the consequences of endowing the gene with these subjectlike powers is that our ideas of freedom and possibility, never that clear, are further muddled. Our freedom seems to be endangered by things that are done to us and affirmed by things we do; the genes as subjects make us objects, just as the behaviorists' stimuli do. Genetic determinism is often criticized for turning us into robots pushed and pulled by biological forces, while the behaviorist emphasis on stimulus control has frequently been denounced for making us passive, merely reactive objects of environmental forces. This two-fronted attack on autonomy should give us pause, especially when we are told that *this is all there is:* just genes and environment. If we are objects of both, then where is our freedom to be subjects? (See also Lewontin, 1983.)

The daunting literature on free will and determinism aside, this problem is central to our attempts to conceptualize psychological freedom, whether for the purpose of assigning moral responsibility or of diagnosing mental "health" (issues that become intimately connected in forensic psychology and psychiatry). Here again we see the vocabulary of subject and object in descriptions of persons driven and compelled by forces beyond their ken, controlling themselves or not, choosing freely or being pushed by instincts, drives, or compulsions. Furthermore, presumably owing to our tendency to see ourselves as objects of our biology, whenever a phenomenon is pronounced "biological," because a chemical or lesion has been found somewhere, because something similar has been observed in some other animal or in some very different culture, because some of its variation is heritable in somebody's research, because a gene "for" it has been found or postulated, because a center "for" it has been found in the brain, freedom and responsibility seem eroded. Sometimes this is seen as a boon, as in the case of an alcoholic who learns with relief that what he or she "has" is a disease, possibly even a genetic one. This releases the person from the burdens of responsibility and guilt, and, paradoxically and most important, in many cases allows him or her to take responsibility for this behavior for the first time and to change it. Sometimes, on the contrary, a claim of biological causation is an insult or a threat, as happens when a proud homosexual is told that his or her sexual preference is an illness, or when a feminist is told that her arguments are a function of the time of the month. We often seem, that is, to welcome biological explanation for things for which we have difficulty taking responsibility, while we resent it when it explains away our choices. Designating persons as "victims" or "products" of their environments involves similar complexities, conceptual, political, and moral.[4] On the basis of such causal assumptions we may forgo action because failure seems assured before the fact; we may convey our expectations of the inevitable to beings exquisitely ready to incorporate those expectations into their conceptions of themselves, thus increasing the probability that the prediction will be accurate; or we may conduct ourselves individually and collectively in ways that we believe universal and decreed by the very nature of life, and in so doing fail to seek alternative ways of living tolerably together. On the other hand, we may, by seeing persons simply as products of conditions, undermine the very capacities for self-determination that are necessary to alter those conditions. By accepting a simple-minded notion of learning, we may assume that because someone does not respond to some fairly short-lived, trivial, or gross aspect of the environment, he or she is beyond influence. Or we may believe that whatever is learned can be changed. We do equal violence to the complexity of human life, that is, by overestimating its rigidity and by underestimating the subtlety and structured conservatism of its modes of change. We do most violence of all by seeing persons and other organisms as mere effects of genetic and environmental causes rather than as active beings that to some extent define their own possibilities. These are some of the reasons why it is important to clarify our ways of thinking about causation; as will be detailed in the next section, the consequences of our metaphors and models reach far beyond the scientific enterprise in which they are born.

While it may be the case that drawing binary distinctions is a fundamental step in knowing (including the process of individuating and knowing ourselves), and while the inside-outside, self-other, subject-object relations may necessarily undergird many of our mental processes, our view of ourselves as sources and recipients of causal influence need not be limited to these distinctions. What I am suggesting is that much more is needed than a modification of some technical vocabulary to set our conceptual house in order, for that vocabulary is too intimately connected to the rest of our conceptions of ourselves and the world to be altered in isolation. When such fragmentary terminological corrections are attempted, as has often occurred in the areas under discussion, they therefore accomplish little. It matters not very much whether knowledge, causation, and control are conceptualized in terms of our own activities or in terms of machines or other entities made in our image; the ideas of action and responsibility are basically the same. Giving up the myriad versions of a sterile nature-nurture distinction, then, requires giving up our traditional notions of simple cause and effect, of subject and object as mutually exclusive opposites, and of order as translation or imposition rather than as transformation and emergence.[5] Living order is not simply a reading out of preexisting order; it is not composed of portions that actively form themselves from a pattern and portions that are passively formed from without. Nor is there a hierarchy of causes, some that truly instill form and others that are somehow not really causes at all, but only necessary conditions or raw materials. The biological, the psychological, the social, and the cultural are related not as alternative causes but as levels of analysis. Whether control is internal or external becomes a matter of point of view, and psychoanalysts have long known that it is possible to control by passivity.[6]

C. Universe as Message

In addition to the concern over the extent to which ontogeny and adaptedness are results of quasi-intentional genetic action, and related to this concern, other preoccupations are detectable in the various versions of the nature-nurture debate. They are still connected to genes and environment as alternative causes, but they are removed a step from questions about how things actually happen in life cycles. Instead, they are implicational. What, if anything, is inevitable? What limits exist to our developmental, social, cultural, political, and moral possibilities? What are the consequences of exceeding proper control over our own and others' destinies? How are the normal, the natural, the desirable to be conceived, and what is the relationship among them? Is there some kind of more fundamental reality underlying the flux of the observed living world?

These are solemn questions. Adequate respect for them requires treating them separately rather than as aspects of a single, more general question that is itself unintelligible, for such lumping together only hinders our efforts at responsible deliberation. Our beliefs about precisely what constitutes relevant evidence on them need scrutiny; we are certainly not well served by the reflex assumption that the biological in any of its guises is a guide.

We have thought for centuries that, by studying nature, we might come to know more about God's nature and will, and thus about ourselves. The Renaissance doctrine of signatures taught that nature could be read as a network of symbols; today we read primate, human infant, or hunter-gatherer behavior for signs about our own aggressiveness, roles, and abilities. In addition to being its marvelous self, the universe is seen as containing messages to direct us in our moral and political wanderings. This is why the logical, methodological, and terminological problems catalogued in the preceding pages are also something more than technical issues in the sciences; they are definitive demonstrations that this diviner's labor is misguided, futile, and even perilous. It is doomed at its start because the distinctions it is based on, between genetic and environmental forms, innate and acquired traits, are not viable. It is doomed again because what is, is not necessarily what must be, not only in the sense that past risings of the sun fail to guarantee tomorrow's, but because what must be in the biological world has been erroneously identified with the genes rather than with actual causal interactions, whose necessity is defined by those interactions. It is doomed next because what is, is not necessarily what should be, and yet again because the limits and price of control must be empirically investigated and evaluated according to explicit values; they cannot be illuminated by designating traits, processes, or variation environmental or genetic. Finally it is doomed because biological reality is not a kind of Platonic ideal imperfectly glimpsed in the phenotype. These questions about inevitability, desirability, and reality, which we will consider separately, mix the philosophical with the scientific in peculiar ways. It is not that these modes of inquiry, so recently distinguishable, must be strictly divorced; but in a society that gives quick credence to scientific fact while deeming philosophical judgment as multiple, indeterminate, and irrelevant to real life, it is important to identify, as far as possible, observation, inference, and value, and to be very careful about their relations. Conceptual concerns like those that make up the body of this book are often dismissed by scientists as being mere philosophy, not science (and not even theory, because they generate no specific hypotheses), but if what constitute data and what those data mean are not legitimate scientific concerns, then what are they?

As was the case with types of genetic metaphors, these issues are interrelated and overlapped. I separate them artificially for simplicity, but they should be cross-referenced by the reader. We will look first at the idea that the genes define the inevitable, then at the belief that they tell us what is normal or desirable. Next the concept of genetic limits will be examined; it is closely tied to the first two. Finally the assumption that we can discern basic reality by reading nature is discussed; it has several common versions, which share the premise that there are ways of penetrating the veil of phenotypic variability and of discerning behind it the fundamental forms given by the genetic code. Along with the considerations outlined in the first part of this chapter, they represent some of the ghosts in the conceptual machine that so reliably produces genetic ghosts in living machines.

In the pages that follow, I will examine several books in some detail. Since it is too easy (though tempting) to select and display outrages from the worst the literature has to offer, or to do violence to an author's overall intention by picking a few nits per book, I have chosen, in addition to briefer examples from a variety of sources, three books for more detailed treatment. All three authors—Melvin Konner, Mary Midgley, and Harold Fishbein—are reasonable, humane, and learned in many respects, and I have tried to give a flavor of their approaches to the biological.

I. THE INEVITABLE

In its crudest form, this is the belief that genetic traits are inevitable, impervious to all influence. If a "genetic" disease or deformity is at issue, it must be accepted as an act of God. Though few scientists are so naive as to hold this belief (but see note 7, Chapter 8), they commonly hold one or more of its variants. Having explained that genes don't preordain but do predispose, for example, some authors reify developmental canalization or epigenetic rules for traits in the actual world and draw conclusions about their distribution in other worlds. Fishbein declares, "what distinguishes canalized from noncanalized acquired behavior, then, is *not* whether it is learned or unlearned, but rather the ease and inevitability with which it is learned" (1976, p. 116).

Konner, whose 1982 book *The Tangled Wing* is subtitled *Biological Constraints on the Human Spirit,* addresses the issue of inevitability. He gives several enlightened discussions of the nature-nurture dichotomy, citing with approval Anastasi's well-known declaration that both dichotomous and quantitative formulations of hereditary and environmental influences are nonsense, and that the only valid question about development is *how* it occurs. Konner characterizes the metabolic disorder phenylketonuria (PKU) as totally genetic and totally environmental because it is traceable to a single gene but is caused by an "environmental poison," and concludes that it is "folly" to "partition mental and behavioral characteristics into percentages of genetic and environmental causation." [7] He further points out that knowing genetic causes allows us to find environmental cures (pp. 80, 89, 104; all subsequent references are to this volume).

In a chapter called "Rage," he declares, "It is obvious that I believe in the existence of innate aggressive tendencies in humans," but he says he will refrain from citing violent societies, which would be easy but would not serve his main goal: the destruction of the "myth" that there exist societies "in which people are incapable of violence" (pp. 203–204). He goes on to describe Dentan's work on the Semai of Malaysia, a nonviolent people used in battle by the British to combat Communists in the 1950s. On the battlefield these usually gentle people were reported to kill violently and sometimes even to enter a kind of blood-trance. From this, Konner concludes that the Semai, in spite of their peaceable socialization and tranquil ways, retained the capacity for violence: no cultural training, he says, can eliminate this "basic core of capability of violence that is part of the makeup of human beings" (pp. 204–207).[8]

Because this seems a measured, modest conclusion, based on ethnography, not fantasy, and because the author clearly takes no pleasure in it, having made evident his distaste for violence and having stated that explaining aggressive behavior is not equivalent to condoning it (p. 180), there seems to be little to object to. But look at the comments that close the chapter. "The continued pretense by some social scientists and philosophers that human beings are basically peaceable has so far evidently prevented little of human violence—which latter achievement would be the only possible justification for its benighted concealment of the truth" (p. 207).

Some have certainly argued that humans are basically peaceable; this is not the same as claiming that humans who live in peaceful societies are "incapable of violence," which is the "myth" Konner sets out to abolish. I am not aware of anyone who believes this latter myth, especially if the humans in question are removed from their native culture and impressed into battle service. As evidence against his claim that humans are basically aggressive, Konner seems to require a demonstration that there exist humans who will not be violent under *any* circumstances. He does this to refute the claim that humans are basically pacific. But by his own criterion, this latter refutation would require showing that there are humans who will not be *peaceful* under any circumstances.

If the admittedly dismaying behavior of some Semai in combat proves that humans are not basically peaceful and that they possess an ineradicable core of aggressiveness, is it also true that, because even people in violent societies sometimes seek peace, we are not basically aggressive either? That, indeed, we possess an indestructible core of capability for peace? Should we argue that any activity that can be evoked under the proper circumstances (and, perhaps, to be more faithful to the example of the Semai, that is then engaged in with some spontaneity or passion) is an innate tendency, a "basic core of capability" that may be "reduced, dormant, yes," but is "never abolished," "never nonexistent," "always there" (p. 206)? One could say this, of course, though it would be an admittedly odd way of saying that we "possess" the capacities to do whatever we are capable of doing, if adequate conditions arise. But this is surely not what Konner intends. Again and again he distinguishes the genetically encoded from the nongenetically transmitted (pp. 20, 22), the inherited from the acquired,[9] those aggressive feelings in which "genetically determined fixed action patterns and releasing mechanisms play a powerful role" from those in which they play little or no role "with the innate factors reduced to some characteristics of motive and mood" (p. 186). These are not trivial lapses in terminology, despite the author's explicit ridicule of the nature-nurture dichotomy and the partitioning of variance (Anastasi's first and second questions: Which? and How much? [1958]). They are what allow him, even oblige him, to speak of some traits as "in" us, whether we display them or not. They are the foundation of the diviner's approach to biology, psychology, and anthropology, and they are part and parcel of the view that leads him to align himself with biologists, who, like the creators of literature, possess a dark vision of "the unchanging facts of human nature," and against social and behavioral scientists who espouse what he calls a "tinker theory" that anything can be fixed by changing institutions or techniques (pp. 413–417).

Konner is quite right in rejecting the naive belief that anything and everything can and should be fixed. But why should this belief be less compatible with a biological perspective than with any other, unless the biological implies fixity and inevitability? In fact, he identifies as one of the central tenets of ethology the doctrine that "many aspects of the behavior of a species are as fixed as its morphology, and equally as attributable to the genes" (p. 144). (For a discussion of developmental fixity and the genes, see Lehrman, 1970.)

None of the various indices of "biological bases" (heritability of variation within or among species, identifiable chemical or morphological correlates, presence at birth, species-typicality, apparent absence of learning, etc.) necessarily gives information on whether a trait is avoidable or whether it is modifiable once present, which is not the same thing. Nor does a conviction that most behavior is learned or "culturally conditioned," a conviction traditionally associated with social science, necessarily lead to the tinkerer's mentality. One might believe that many learned habits, motives, and beliefs are extremely difficult to influence because they are intricately interrelated in persons or in societies in such a way as to create great inertia, because they organize perception and interpretation in ways that discourage disconfirmation, because satisfactory alternatives are not available, because social change is possible but undesirable given prevailing economic and political realities, and so on. What explains this fairly predictable choosing of sides and positions, of course, is the nature-nurture tradition, with which Konner is very comfortable, his protestations and sophistication notwithstanding. Though he has some pertinent comments to make on the unfairness of labeling all biologists political reactionaries, and though he rightly points out that a given belief about biological fixity could be used to justify politically progressive policies as well as conservative ones, the biological givens in his examples (gender differences in aggressiveness, individual differences in intelligence and character) are instructive (pp. 418–421). Though individual differences (and gender differences—Konner is cautious about other group differences) are multiply caused, he says, it is "naive" to deny that the genes can be a source of differences among humans. "There is in each of us a residue of characteristics of heart and mind that we brought with us when we entered the womb" (p. 105).

I agree it is useless and misguided to deny that genes can be a source of differences among individuals and among groups. It is unfortunate that philosophy, history, and some of psychological theory have arranged for a belief in the importance of experience as a source of differences and in the importance of learning, which is not the same thing, to be associated with this denial. But genes as sources of variation in particular situations is not the same thing as characteristics of heart and of mind present in the womb, because just what aspect of what character a gene may affect, whether it has an effect at all and how, is not always determined at conception. The relation, that is, between genes as sources of differences among individuals and genes as *fixers* of behavior is obscure. Obscure, that is, unless one subscribes to the whole complex of beliefs about form, origins, and order that have been the subject of the present critique.

Having concurred with Noam Chomsky's ideas on the innateness of language, Konner says that the environment is necessary and important in language development, but that it is "of small *formative* importance." A skeleton, he continues, has a genetically determined structure that develops maturationally (quite rigidly). It needs nutrients or the plan goes "awry," producing deformities that may be so severe as to render the structure nearly unrecognizable. Many other things, including number of calories, stress, light, and exercise, influence growth rates and resulting structure. But "any knowledgeable scientist" agrees that the basic structure is produced "under the aegis of the genes" in the normal environment (pp. 156–157). The reader will here recognize the standard view, expressed by Monod, Gesell, and many, many others who have offered opinions on the relevance of biology to our lives. The genes carry the form of innate characters; they are not seriously subject to outside influence, so they define the givens of our private and public existence, past, present, and future.

That control flows only from the genes outward, never back, is dogma for many. The failure of Lamarckianism and of instructional theories in immunology seems to render that dogma unassailable (Barash, 1977, p. 28; Jacob, 1982, pp. 12–18). Jacob considers the popularity of instructional theories to derive from our habit of projecting human mental processes into biological processes. He does not comment on the anthropomorphism of theories based on genetic programs, which, in fact, he sees as the antidote to anthropomorphism. But the world need not construct or alter the genes in order for "information" or "control" to flow inward. When "genetic information" is expressed or "genetic control" exerted, this usually means that a selection is made from a number of alternatives—

this happens, not *that*. Similarly environmental control of genetic processes can simply be a matter of selection from among genetic possibilities. The crucial distinction here is between environmental "influence" as actual alteration of genetic structure and as control of genetic activity and impact. If information could not flow inward in this latter sense, development would be impossible.

Since the genes are involved in transient changes and major sequences alike, in both species-typical and unusual processes and products, how can the basic or inevitable be defined with reference to them?

2. THE NATURAL, THE NORMAL, THE DESIRABLE

The relation between fact and value, "is" and "ought," has troubled us for many generations. Scientists tend to ignore it, considering it nonexistent by definition, or to treat it in a gingerly fashion. It is difficult to avoid the issue, though, and it is interesting to look at what happens when writers interested in biological matters confront it, however obliquely.

E. O. Wilson (1978), in a book that one suspects was written at least in part to counter the charges of reactionary biology referred to by Konner (see also Barash, 1977, pp. 278–280, who dismisses this kind of criticism of sociobiology by saying that the biological is not always good and that ethical judgments are out of place in science), states that Christian natural law is an erroneous "theory" on biological laws, written by men quite ignorant of biology. He goes on to say that "human genetic history" argues for a more liberal sexual morality, in which sexual practices are to be regarded first as bonding devices and only second as means for procreation. One of the worst examples of the "sanctification of premature biological hypothesis" [10] is the labeling of homosexuality as unnatural and disordered. In fact, Wilson argues, homosexuality may be "normal in the biological sense . . . a distinctive beneficent behavior that evolved as an important element of early human social organization. Homosexuals may be the genetic carriers of some of mankind's rare altruistic impulses" (Wilson, 1978, pp. 141–143).

Wilson goes on to examine the policy implications of this postulated adaptiveness by altruism. (The argument is that homosexuals might benefit kin, perhaps by remaining at home to help rear children, or by providing other advantages to them. No evidence is offered for this, and one could even argue that the sexually indifferent, for instance, might be more likely to stay at home or aid the family than homosexuals, who, after all, are defined at least partly by their propensity to form sexual attachments.) It would be wrong, he says, to consider homosexuals "a separate genetic caste." It would be even worse to make "past genetic adaptedness" a condition for acceptance today. Worst of all, though, is discrimination on the "unlikely assumption that they are biologically unnatural" (p. 147).

Though I absolutely agree with Wilson on the undesirability of discriminating against homosexuals, I find it difficult to follow his reasoning to this position. He refuses to make "biological normality" a condition for acceptance. If at least some things that are not normal in his sense can therefore be accepted, and if at least some things that *are* normal can be disapproved of,[11] then what is the relation between biology and morality? Why should it matter at all whether homosexuality or anything else can be hypothesized to be natural or unnatural by his definition? How would one demonstrate unnaturalness? By showing lack of contribution to inclusive fitness?[12]

Perhaps what Wilson really wants to say is "Just because I am a sociobiologist, I don't necessarily condemn people who have atypical sexual preferences or who don't reproduce." (Many homosexuals do reproduce, of course, and many heterosexuals do not.) If so, one might wish he had done so and left it at that. But his whole point is not that biologists may separate their scientific constructs from their morals, but that the latter are partially explained and justified by the former. And thus the muddle. (See Mattern, 1978.)

Even Konner, who is fairly clear in his refusal to derive the good from the natural, and who frequently justifies the "biological" approach as our only hope of curbing, modifying, or otherwise coping with the undesirable in our nature, suggests that we may learn from hunter-gatherers, who he believes represent quite accurately the "human condition as it was lived when people were evolving," that the "ethic of sharing is not 'unnatural.' We may thus try to revive it even though we dwell among nonrelatives" (1982, p. 10). Though biology cannot dictate our values, it can apparently be used to justify values we wish to advance or to estimate the probability of advancing them successfully. But Konner acknowledges that hunter-gatherer social systems are different from ours, that we cannot "go back" to theirs; it would therefore seem that having their precedent doesn't really increase our chances of nurturing an ethic of sharing in the modern urban world, unless it can be demonstrated that we are similar to them not only in genetic constitution but in those other aspects of our developmental systems that are necessary for producing and maintaining such behavior. Nor does it make that ethic more desirable or justifiable.

Our world has never existed before, and its conditions are constantly changing; what is new exists as certainly — and, with hindsight, as "inevitably" — as what is old. Is it less likely that we could develop a set of institutions and values absent among hunter-gatherers than that we could "revive" one that is present among them? (If so, the modern world seems a highly improbable development.) Not unless one endows hunter-gatherer qualities with a power and significance that transcend physical, cultural, and social environments, which is exactly what the conventional biological view does, despite pronouncements that no trait appears without the proper environment, that genes and environment are both important, that nature-nurture is nonsense, and that all is "interaction." According to this view, what is genetic is in us whether it is manifested or not, in a way that the environmental or acquired is not. Specific conditions may be needed to *release* it, but not to *create* it.

One might argue, incidentally, that an ethic of sharing does exist in modern society, in the form of taxation, charities, admiration of generosity (including that of preliterate peoples), disapproval of greed, socialization emphasizing sharing, and so on, and that these all coexist with values and institutions that exalt, instill, and reward rapaciousness. I suspect that it is just this situation that accounts for our rather wistful backward glances toward apparently simpler, kinder times. One wonders, though, whether it was ever easier to balance ambition with charity, greed with compassion, eros with agape.

A philosopher, not a biologist or social scientist, Mary Midgley ponders the relation between fact and value in her book *Beast and Man* (1980; all references are to this volume) and embraces a biologically based morality without equivocation. She distinguishes the "strong sense" of natural as involving approval or recommendation from the "weak" sense, which is simply a statement that something exists. Things that are natural in the latter sense can also, she says, be called unnatural and undesirable. On the example of sadism she asserts, "that consenting adults should bite each other in bed is in all senses natural; that school teachers should bully children for their sexual gratification is not" (p. 79). The structure of the instincts "indicates the good and bad for us." If people are naturally inquiring, then inquiry is an important good for them; they should do it, and they should not attempt to prevent each other from engaging in it "unless they have to" (pp. 74–75). Again, a species's basic wants and needs are given by its nature, and "if we say something is good or bad for human beings, we must take our species's actual needs and wants as facts, as something given. . . . It is hard to see what would be meant by calling good something that is not in any way wanted or needed by any living creature" (p. 182). (One would have thought that the problem was rather that of calling "good" only some of the things some creatures want.) Our conscience is but our own nature "becoming aware of its own underlying pattern" (p. 274).

Midgley's argument involves more complexity than can be treated here, and she has important points to make about the symbolic significance of animals, as well as about the traditional opposition of reason to feeling and desire.[13] Unfortunately she uncritically adopts virtually the entire empiricist-nativist conceptual apparatus, from innate instincts, which she divides into closed and open (pp. 52–53), to the assumption that instincts are the only alternative to the tabula rasa. She defines "entirely acquired" as "totally arbitrary and unrestrained, infinitely variable and docile" (p. 287) and dedicates the book to her sons, whom she thanks for showing her that human infants are not "blank paper." In her introduction she says that if we have no nature, we must believe that we are *"infinitely"* easy to indoctrinate, since there could be "no inborn tendencies that either deflect or resist the process" (pp. xx–xxi). She does not consider the possibility that our natures are contingent, changing, and multiple, but not infinitely so, that the environment is itself structured but changing, and that it, too, imposes limits on variation. Significantly enough, though she quite rightly says that it makes no sense to oppose culture to nature for groups or for the human race as a whole, she accepts that opposition for individuals (never explaining how she accomplishes this [p. 285]). She thus retains the framework that allows her to assume a single genetically determined set of wants, needs, and propensities on which to base an ethical system.

Just as she assumes a unity of human nature, she implies a single static world. To feel at home in the world, she says, we don't have to believe that it was made for us, but vice versa. "We are at home in this world *be-*

cause we were made for it. We have developed here, on this planet, and are adapted to live here. Our emotional constitution is part of that adaptation. We are not fit to live anywhere else." This truth is not challenged by our ability to survive briefly and "at ruinous expense" in peculiar environments like spacecrafts, since this only shows our ability to transfer our environment to odd places, not our ability to live in a different one (p. 195).

The fact is that adaptation is not itself unitary or static and is only meaningful with respect to a particular environment. We live in *many* worlds, all different from the ones that existed many thousands of years ago, and each world, even if reduced to the unique world of the individual, is not unitary. It is composed of roles and settings that may overlap, conflict, and impinge on other complex worlds; it changes, sometimes rapidly and not necessarily in synchrony with others. Surely the point of any ethical system is to aid us through this labyrinth, something Midgley herself implies when she speaks of conflicting wants and needs. Even if there were justification for postulating a single set of human needs and wants, that is, it is not at all clear that it could serve as a foundation for an adequate ethical code: Moral decisions often become necessary when several people want the same thing but not all can have it. Fortunately or unfortunately, depending on one's point of view, there is no justification for such a postulation. To assemble a catalogue of basic needs requires one to assume precisely what is supposedly being defined: fundamental values that can order incompatible and unequal considerations.

Individuals and cultures cannot create or annihilate wants, Midgley asserts, though they may "group, reflect, guide, channel, and develop" them. (Notice the resemblance to the conventional description of environmental influences on gene expression.) The wants all people have in common are, and this is presumably not an exhaustive list, "to move fast, to do their business quickly, to be honored, feared, and admired, to solve puzzles, and to have something bright and shiny" (pp. 182–183). But when a sadistic schoolmaster's wish to be feared (and perhaps to enjoy a sexual thrill to boot) conflicts with his pupils' wish to avoid pain and humiliation (basic wants?), Midgley has no hesitation in denouncing the educator's wish as unnatural, not only because it inflicts injury, but because it is off center. An "unnatural life" is one whose "center has been misplaced"; its parts have taken over and have been allowed to "ruin the shape" of the whole

(p. 80). Values of internal harmony, moderation, and balance as well as a disapproval of obsession seem to guide her decision here. Yet they do not figure in her brief catalogue of universal wants, and would be difficult to justify as universals or even as predominant in our own society, whose heroes are often single-minded, off center, and obsessive in the pursuit of some good, or even of some evil (but see her pp. 192–193).

Midgley declares a unity of nature, not only in the sense of shared needs, but also in the sense of a shared need for integration, a convergence of many goods toward one. She deals with the facts of multiplicity and conflict of values, however, by utilizing higher-level principles that are not found in her roster of universals, buttressing her claim for primacy of balance and proportion by reference to the ethological concept of supernormal stimuli, which can lead to maladaptive, exaggerated behavior. *"Obsessiveness unbalances people's tastes against biological advantage,"* she says, arguing not for a bourgeois or conventional notion of balance, but for a natural one (pp. 190–194, 264–265). But in spite of the carefulness of her presentation, one fears that the unity is artificial. Supernormal stimuli are usually an intrusion into a creature's normal environment (an enormous egg placed in a nest by a curious ethologist, for instance, may seem more attractive to a bird than a normal egg); speaking of human behavior in these terms would seem to require considering some of our worlds unnatural. Midgley considers foot binding and the encouragement of obesity in the service of feminine beauty travesties of the natural. The idea of universal "biological advantage" that is diminished by such customs is obscure, however, especially given the traditional evaluation of advantage or fitness by differential reproduction and the roles of bound feet and fatting houses in the courtship systems of their respective cultures. Fitness is not absolute but relative, to other reproductive rates and to a setting, and no idea of adaptiveness makes sense without a niche, a "world," that is adapted *to*. The sense of biological advantage that Midgley intends, then, seems not to be that of evolutionary biology after all, but a conception of the good life that has no clear relation to reproductive rates.

Having unified diversity of nature by invoking a higher good toward which the many converge (a move that she seems to imply is justified by the alleged universality of the desire for integration, but which does not in fact follow even if that universal is granted), Midgley does the same for worlds, the multiplicity and defectiveness of which are everywhere evident in her descriptions. We were *not* made for this world, after all, or at least not for all aspects of it, and whatever criterion Midgley employs to discriminate among subworlds and their parts is neither universal nor derivable from universals.

Midgley's efforts are not examined here because her approach dominates her field. Indeed, my impression is that philosophers tend to be as reluctant explicitly to derive morals from facts as are most scientists (for some discussion, see papers in Caplan, 1978, and Stent, 1980). Rather her work is of interest because, on the one hand, it shows how far-reaching and pervasive are the effects of a misconceived nature-nurture dichotomy, and, on the other, it allows us to see, clearly and in the open, what is mostly a hidden agenda for other writers. Since Midgley acknowledges that she is reasoning from facts to values, there is less need to ferret out implicit assumptions and covert logical operations than there is with authors who disclaim any such endeavor. The advantage to having both approaches to compare is that their essential similarity becomes clear once one takes the trouble to go beyond the writers' stated intentions and to look at what it is that they actually say.

3. LIMITS

The notion of genetic limits is in many ways a combination of the first two items in this review of questions whose answers are often sought in biology. One might not think that a trait was inevitable, for instance, but still believe that preventing it or modifying it was too costly or difficult to warrant the effort. The reckoning could be done in the currency of time, human stress or pathology, technical resources, number of related changes entailed, and so on. Or one might not think that a "natural" trait was desirable, but consider it preferable to the costs of trying to change it (see previous example) or conclude that, because it was a given of human life, it should be accepted or accommodated to.

Midgley comments that Plato's Republic required violation of natural tendencies and that its implementation entailed too great a distortion of natural sentiments to be viable. She concedes that such sacrifices may be made, in religious or communal groups, but insists "they *are* sacrifices," paid for with great effort, vigilance, and the danger of "emotional stunting and self-deception" (pp. 75–76). To pronounce an activity natural, she says, is not just to approve of it, but to comment on the sorts of advantages it confers, "and also the *sorts* of dangers likely to attend forbidding or neglecting it. If it is natural, in the strong sense, it fills a need, and one that cannot easily be filled with a substitute" (p. 186).

Konner concludes that we cannot entirely suppress or eliminate the genetic core of characteristics with which the human race is born. It might be possible to reduce or eliminate the difference between men's and women's aggressiveness,[14] for example, by reducing violence in men or increasing it in women. It would be quite difficult to do the former, he believes, while the latter would be "perhaps easier but probably undesirable." In the meantime, given these technological and moral limits to our control, we might do well to place the more peaceful sex in all positions of military and diplomatic power (p. 420).

Toward the end of their book on genes and culture, Lumsden and Wilson say that a society that ignores epigenetic rules is still constrained by them, but that the neglect of these innate factors invites conflict and danger. Trying to escape the influence of epigenetic rules risks losing the "very essence of humanness" (1981, pp. 358–360).

Recall that Fishbein defines canalized behavior by the inevitability and ease with which it is learned. The idea of canalization, and the way it is reified by this writer and others, will be discussed later. For now it is sufficient to note that inevitability, as Fishbein uses it, is essentially a frequency concept. Ease is not necessarily related to frequency, unless it is simply defined in this way—without a metric it is difficult to know what it means. Like many concepts, ease of learning is likely to vary with its operationalizations. The point here is that no criterion has been advanced by these writers, though the implication is that ease and frequency are correlated, if not interdefinable. Though this seems a minor matter, it is not, since the inference of fixity from frequency is one of the basic nature-nurture mistakes; the corollaries to the natural-is-easy doctrine are unnatural (infrequent or not universal)-is-hard and natural-is-hard-to-change, and these are among the most mischievous implications of the notion of genetic limits.

What these treatments have in common is the assumption that what is "genetic" in any sense appears automatically and easily. Modifying such characters while they are developing or after they are formed is seen as taking more coercive intervention, and exacting a heavier toll on the subject in stress and distortion, than does influencing acquired characters. Lehrman's discussion of developmental fixity has already been cited in this respect (1970; see also Block, 1979). As he points out, there is no reason to believe that these assumptions are valid, no matter which definition of the innate is used (and it is important to bear in mind that the various definitions do not necessarily coincide). Characters may be easily altered at some stages of development and not at others, or easy to affect with some techniques and hard with others, difficult to prevent but easy to influence once developed and vice versa, easy to change in one direction but not in another, and so on. What is technically easy may entail much effort or stress on the part of the subject, and effort and stress are not the same thing. It is possible to postulate a species-typical, "biological" sequence that is long, arduous, and stressful, full of psychic convolution and conflict. Freud's psychosexual stages constituted just such a sequence, and so do everyday notions of psychological and moral maturity. When age brings wisdom, it does so "naturally" but not necessarily simply. Dinnerstein (1977) gives an account of gender development as complex, fraught with bafflement and pain, anger and disappointment. She considers it universal because it depends on universal child-rearing arrangements.

Many behavioral anomalies develop as automatically and smoothly (or bumpily) as do the common patterns, and the price of any behavior, common or rare, as well as the price of changing it, depends on the world in which it is exhibited. Finally, we can refer back to the discussion of constraints, and recall that the limits to control depend on what one is attempting to control, when, and in what manner. Like constraints, limits have an ontogenetic history and must be considered in that context.

4. BASICS

The fourth question often put to nature is, What is real, basic, necessary, fundamental? Frequently it is answered by pointing to the genetic in one or another of its guises. What concerns me here is more elusive and subtle. It seems to occur when an investigator eliminates the easy answer of "the innate" by conceptual analysis (or increasingly, as conventional interactionism becomes obligatory, by adherence to a new dogma) but has apparently failed to convince him- or herself that there is not still a basic hereditary *something* hidden in the data, if only the right formula can be found. Sometimes the method of approaching this hidden reality is indirection: One talks around it, looks just to the side of it, as at a dim star, or tries to define it by chipping away at what it is not, in the hope that its latent outline will be revealed.

Before my own descriptions become even more convoluted, let me turn to some examples. They involve the search for (a) special kinds of development, (b) categories of behavior, and (c) cores of personality.

a. The basic as special developmental processes. Harold Fishbein, in a book called *Evolution, Development, and Children's Learning,* describes evolution and learning as "two major processes" for population adaptation, establishing at the very beginning the notion of alternative sources of information that was seen to underlie the nature-nurture antithesis (1976, p. 1; all references are to this volume). An avowed "interactionist," [15] he nevertheless speaks freely of encoded designs, of "hereditary mechanisms contained in the genotype," of knowledge contained in the genes or acquired through the phenotype. These notions he attributes to Waddington, Lorenz, and Piaget, a rather motley trio he invokes repeatedly, especially with reference to canalized development, defined as "genetically preprogrammed" development (pp. 5–9, 57). (Only Lorenz's thought is actually consistent with this conception of psychological development. Though Piaget to some extent accepted genes and experience as alternative sources of information, he certainly did not view cognitive development as genetically programmed, no matter how regular and predictable.)

The central theme in Fishbein's book, the notion of canalization, deserves special attention. Though he attributes the idea to the authors named above, it is found specifically in the work of C. H. Waddington, who consistently refused the nature-nurture distinction, and who used canalization to refer to developmental regulation—the tendency for similar results to appear despite some variation in developmental conditions. Canalization is, as he repeatedly said, a matter of balance of flexibility and inflexibility in development, and it was initially proposed as a way of describing the appearance of a limited number of fairly well defined tissue types in differentiation rather than continuously graded series (1975). The tendency to regulate in spite of particular kinds of environmental or genetic variation varies with species. Though the term "regulation" tempts many to think of development as a process regulated by intelligent genes, Waddington did not regard it as a function of the genes alone, instead seeing it as "a developmental process whose course is steered by the combined action of the whole genotype and the impinging environment." Abnormal or atypical phenotypes can also be the result of canalization; all this means is that "there are a considerable variety of genetic changes ('mimic genes') and of environmental stresses, any one of which will steer the development toward the production of a particular phenotypic abnormality" (1975, p. 77). This is crucial to his discussion of phenocopies and genetic assimilation.

All this should make clear that Waddington would not have approved of Fishbein's association of canalization with designs encoded in genes; with images, templates, or phenotypic targets contained in them; or its unqualified identification with the adaptive (1976, pp. 5-9). In fact, much of his work is an attempt to offer an alternative to this way of thinking. Canalization, as Waddington conceived it, is demonstrated when some phenotypic feature is resistant to variation in genes, environment, or both. Fishbein does mention research in which canalization, as resistance to selective breeding, was investigated (pp. 35-37; this is interesting because it makes degree of canalization *inversely* related to heritability, if the latter is defined as response to selection). A developmental system that is canalized with respect to variations in temperature may not be equally buffered against variation in humidity, and genetic changes may have no, little, or great impact on development, depending on the rest of the genotype and on surrounding conditions. Though Waddington occasionally generalizes about canalization and developmental systems, especially when speaking about the importance of phenotypes and epigenetic processes in evolution, he insists it be demonstrated with respect to particular variation and particular systems. He resists inferring it from other phenomena or inferring canalization against one kind of stress from lack of response to another. He even describes very different kinds of stress-response curves that might be found in different cases (1975). Depending on the process, the organism, and the ecological niche occupied, different kinds and degrees of canalization may be adaptive or maladaptive, and adaptive traits may or may not show canalization.

Though he seems to adopt Waddington's thinking, Fishbein actually adopts a *term* and uses it in a manner quite at odds with Waddington's views. He alternately *defines* canalization by species-typical development in the normal environment (pp. 38, 238) and *infers* or *assumes* it from constancy under normal conditions (pp. 9, 18, 44–45, 138). He further assumes that our epigenetic system is that of the primitive hunter-gatherer, and identifies this system with the genes, which we presumably share with those groups. According to him, we learn about our own epigenetic system by looking not only at development in our own societies, but at characteristics of past and present hunter-gatherers and our phylogenetic relatives, who are assumed to share it. Rather than using canalization to illuminate the balance of stability and flexibility in developmental systems by analyzing phenotypic responsiveness to particular influences, Fishbein seems to be employing it as a cover term for the genetic, the adaptive, the natural, the inherited. Waddington's notion of epigenetic systems and canalization as joint functions of genes and environment becomes, in Fishbein's thinking, genetically preprogrammed development, "a very active process in which genes compensate for, or collaborate with one another to insure that phenotypic development will reach those targets" (p. 35). One couldn't wish for a better description of the homunculoid gene.

Fishbein reports Kummer's research on the hamadryas baboon, which normally forms single-male "harems" in which the females closely follow the male; and on the anubis, which forms more loosely organized groups. When placed in a hamadryas group, anubis females took less than an hour to respond appropriately (by following) to hamadryas male threats. Their scores for following, according to Kummer (1971, p. 100), equaled those of control hamadryas females introduced into the same group, though they escaped their males more frequently. From this Fishbein concludes that behavior canalized in the hamadryas females, who normally follow males, is not canalized in the anubis, though the latter can learn it (p. 117).

This is a puzzling conclusion. Though the anubis females showed that they can adjust to a tightly organized routine that is quite abnormal for their species, this tells us nothing about canalization of following behavior by Fishbein's primary definition, inevitability in the *normal* environment. Nor does it tell us anything about canalization toward the abnormal (for the anubis) docile phenotype, though the fact that these females adjusted shows that such canalization is at least possible.

Fishbein cites no evidence for his declaration that following in the hamadryas females is resistant to specific genetic or environmental variation; he seems here to be obeying his stated rule and *assuming* canalization from appearance in the normal environment. In fact, Kummer's (1971, p. 100) observations were that hamadryas females do not follow males if placed in a group of anubis, but instead begin to roam and socialize freely. This is evidence *against* canalization as Waddington defined it. In both cases of female transferral it is responsiveness to conditions, not independence from them, that is demonstrated.

Fishbein's point, I suspect, is simply that responding to male threat by following is more normal for the hamadryas female than for the anubis (which we already knew, since "harems" are typical of the former and not the latter). That this is the case, and that Fishbein's conception of canalization is essentially nothing more than a standard "nature" construct in search of acceptable expression, is evidenced by his observation that socialization among certain catarrhines (humans, gorillas, chimpanzees, savanna baboons, hamadryas baboons, and macaques) is very similar, that it "appears to be canalized" (p. 123). Though Waddington does not, to my knowledge, use canalization to refer to phylogenetic similarity, if we continue our sleuthing and translate "canalized" as "caused by shared genes," the thrust of Fishbein's argument becomes clear. Having noted this cross-species similarity, Fishbein goes on to speculate on what these groups' sexual behavior can tell us about humans. He cites research showing that among certain catarrhines mother-son copulation is rare, while mating between father and daughter and between siblings is common. Seeming to warn against cross-species generalization, he cautions that taboos against the latter kinds of incest may still be canalized in humans. (He has already said that all three taboos are human universals; by his procedural rule they should have been assumed to be canalized.) He then generalizes by saying that Freud appears to have been wrong about the desire among humans for mother-son incest, because such incest does not appear in macaques and chimps.

Since he equates incest taboos, which would seem a peculiarly human phenomenon involving certain types of forbidden conduct, with simple absence or infrequency of behavior in other primates, one wonders why he seems to find the chimpanzee and macaque data more damning to Freud's ideas than the human pattern itself. That is, if relative frequency of various kinds of matings is given equivalent motivational significance in all the species under consideration, then infrequent mother-son copulation in humans surely counts against Freud at least as much as infrequent mother-son copulation in chimps. (Though Kummer does espouse a fairly conventional view of phylogenetic programs and their ontogenetic modification, he is rather more circumspect than some of his interpreters about cross-species inferences, including inferences about morals [1971, 1980].)

The problem, of course, comes from impoverishing the concept of taboo, which has to do not with frequency of events per se but with their meaning. But this is precisely the kind of impoverishment that is necessary to this kind of "biological" reasoning. There is a persistent playing with levels of analysis in such treatments. Distinctions between the level of individual motivation and that of institutions or customs are ignored as the social, the cultural, and ultimately the psychological are collapsed to the biological. At the same time, a new level is created—a phantom plane of genetic reality, which is not observed as such but deduced (or assumed). From this perspective, the observation of living, breathing animal-machines is of value only insofar as it gives access to the ghostly forms and causal agencies within them.

When Fishbein considers sex roles among hunter-gatherers, he comments that it requires no great leap to assume that the division of labor is canalized, that girls more readily learn the female role, and boys, the male (p. 143). The notion of "learning more readily" requires all other things to be equal. But in the normal hunter-gatherer environment (as in ours), of course, they are not. The statement has little meaning for any society in which males and females are treated, and equally important, are thought about and think about themselves, differently. It amounts to a prediction about the probable outcome of unbiased treatment if it were to occur. What is finally at issue, one feels, is what is usually at issue in such discussions: what is inevitable, what is desirable, what is possible, what is natural. The search for the basic as special developmental processes, then, resolves to the same concerns we have already examined.

The "working assumption" that organizes Fishbein's efforts is that "almost all of the phenotypic constancies or invariances observed in species which reside in a normal range of environments" are canalized. "Normal range" is defined as those "to which the species evolved and adapted" (p. 40). This explains why hunter-gatherers are given more importance in his definition of canalized behavior than agrarian or urban peoples: They are considered to occupy more or less the same ecological niche that our evolutionary ancestors did (not that there is a single hunter-gatherer niche). Normality here is not determined by assessing the range of environments actually inhabited by contemporary and past members of the species, but by applying a historical criterion to select a very small subset of those. Since no significant genetic change is considered to have occurred for forty thousand years, and since canalization is accomplished by active genes supposed to be collaborating to reach goals, it is clear why these small, vanishing societies are favored subjects. The explicit statement of strategy is "to examine the anthropological record of contemporary hunter-gatherer societies, identify and discuss the classes of behavior patterns they share, extrapolate backwards in time concerning man's adaptation, and infer that these behaviors are induced or canalized—man learns them because he was designed to learn them" (pp. 138–139). The same strategy is presumably followed when comparisons among primates are conducted. The last step in this sequence, though Fishbein does not make it explicit, seems to be: identify these behaviors in contemporary populations, thus defining the normal and adaptive for all of us.

Despite the fact that Fishbein devotes much time to behavioral constancy, he is not on the trail of constancy as such, but of genetically preprogrammed constancy. In distinguishing between genuine canalization and constancy produced by "phenotypic plasticity" (p. 40), he posits a version of the inherited-acquired distinction that Waddington avoided and repeatedly criticized.[16]

Had Fishbein addressed the variability of sex roles and sexually dimorphic behavior across cultures (or even within complex cultures), it seems unlikely that these categories would have met his criterion of inevitable, universal, and effortless learning. Most of the myriad aspects of sex-typed behavior vary continuously and widely. Male and female distributions show great variability and overlap, unlike the discrete tissue types that sparked Waddington's imagination decades ago. What the diversity of such behavior suggests is not a developmental world in which a cell becomes either a liver cell or a muscle cell but not something in between, but rather one in which gradations among types are infinite and dimensions are loosely linked, in which an individual cell can be liverlike in gross morphology but slightly more like smooth muscle in enzymatic makeup and more like lymphocytes in motility, and so on. Even if one were to conclude that sex typing does not demonstrate enough uniformity to merit being labeled as canalized, however, this would not tell us whether variation *beyond* the observed range of environments would be accompanied by further phenotypic variation. A finding of considerable uniformity, on the other hand, would not allow simple prediction of outcome in some hypothetical society characterized by a social structure and socialization practices we have not observed and hardly know how to imagine. We are, in fact, still defining, in our own lives and social arrangements, our norms of reaction. This should not be taken as a claim that the behavior of men and women could be indistinguishable in some future world; I have just said that we lack the means to make such a prediction. As we try to use what we know about morphological, hormonal, and psychic differentiation, however, we will be as poorly served by an oversimplified notion of relationships between physiological and social-psychological levels as we are by unreflective extrapolation from existing behavioral patterns, whether human or nonhuman.

Phenotypic plasticity (in the sense of multiple developmental possibilities), or lack of it, is what canalization and genetic assimilation are all about, and these concepts depend on a unified view of development as joint action of genes and chemical environment at the molecular level, organism and environment at a higher level, and interaction at various levels in between as well. In such integrated systems, constancy in interactants is as important to understanding as variation, and types of outcomes may be more constant than their interactants. Fishbein's concern with separating genetically programmed constancy from similarity that is a function of mere phenotypic plasticity, and his concern with attributing resemblances among populations of humans and among primate species to sets of shared genes directing shared epigenetic systems, cannot be reconciled with Waddington's concept of canalization, the "necessary" qualities of which are due to environment-gene complexes, which does not tell us what would happen in any particular alternative world, which can apply to the nonadaptive as well as the adaptive, and which must be described with reference to particular observations of responsiveness of specific developmental systems to specific variations.

Had Fishbein fully adopted Waddington's ideas on ontogeny and phylogeny, he would have been disabused of the convictions that genomes carry messages about our fundamental qualities and that developmental processes can be divided into those that give us access to those messages and those that don't.

If we share characteristics with contemporary or past hunter-gatherers or with other primates,[17] and if sharing is not primarily a matter of excessively permissive definition, as was the case with macaque "incest taboos," this is interesting. It does not necessarily mean that the characteristics are the results of the same developmental processes; nor does it mean that they are more basic to our nature, more necessary, more useful, or more difficult or dangerous to change than characteristics we do not share with them. This is because none of these terms has any real meaning except with respect to a given developmental-ecological setting, and it is just these that to some extent differentiate the groups in question. Shared characteristics and unshared ones certainly do not develop by different principles.

The idea of canalization is reified in these attempts to detect the templates and genetic instructions that create human nature. Though Fishbein has slightly enlarged the traditional idea of maturation by including some learning, his version of canalization is just that: essentially a maturation concept, used in the same way maturation has generally been used, as an attribution of regular developmental sequences to the genes, with the result that they are invested with an aura of rigidity and autonomy (Oyama, 1982). With old-style maturation, the remainder of development was explained by learning; with Fishbein's canalization, it is by "phenotypic plasticity."

The developmental "program" is the functioning organism-environment system. This is so whether the phenotype is rare or common in that generation, and whether it is a successful contributor to the gene pool or dies without progeny. If environment and modal phenotype change together without substantial change in the gene pool, evolution as it is usually construed has not occurred; but the phenotype is still a function of an integrated system and the new complex may be important in preparing the way for a novel evolutionary path. It is possible to imagine a more ample definition of evolution, one in which transgenerationally stable alterations in developmental systems are criterial and which therefore embraces even those sequences of changes that do not involve alterations in the typical genotype.

Quite obviously one can no more explain (in the sense of giving a detailed causal account of) an ontogenetic sequence or outcome by attributing it to a developmental system than by attributing it to demons, maturation, canalization, or genetic targets, or by saying that it was meant to happen, was once useful, or is universal in a species or group of species. Causal explanations require the analysis of relations within the system, the identification of combinations of factors necessary or sufficient for constancy and variation in particular events. This is the usual business of defining "proximate" causes. An explanation by "distal" or "ultimate" cause is a historical one, showing how *this* developmental system and not some other came about; it has no particular implications for the precise causal mechanisms that may be discovered within the system, but it does presuppose a complex of relations, shifting in configuration but showing overall continuity, among developmental interactants. Developmental systems are defined by these configurations, and they have their own dynamics, which determine whether a given genetic or environmental change will have no, little, or massive developmental impact on the phenotype. The *continuity* of these systems by successive life cycles, not the exact *duplication* of a particular set of traits, is what is necessary to evolution. To ask whether a present character is phylogenetically derived is to ask whether its developmental processes have shown substantial stability over evolutionary time; often the relevant evidence on this question, which is a question about the similarity between the present and the past, is unavailable. Even having demonstrated genetic similarity, one would have to show similarity or equivalence of developmental circumstances. Current necessity and utility, which have to do with the relation of some aspects of the present to others and of the present to the future, are quite another matter.

With the increasing popularity of "biological" treatments in psychology, sociology, anthropology, political science, and so on, and with the increasing pressure to avoid simplistic, old-style instinct and innateness explanations, scholars have turned to people like Waddington for guidance in making subtler, more sophisticated use of biology. But some habits are hard to break. Lumsden and Wilson (1981, p. 243) also use canalization as a slightly revised version of traditional maturation. Their use of "epigenetic rules" has been mentioned; they, too, cite Waddington to support a decidedly non-Waddingtonian concept of development. Because these workers begin with strong assumptions about what a biological approach must be, and because these assumptions are rooted in largely unreconstructed ideas of preexisting design, the subtle is coarsened, the carefully drawn distinction is lost, and canalization becomes little more than the operation of the genetic ghost that defines and creates nature.

b. The basic as special categories of behavior. Pursuing a goal similar to Fishbein's, N. L. Munn is interested in detecting the unlearned substrate of experience and behavior. He is, in fact, working in a tradition from which Fishbein's theoretical generation may be said to have descended. He begins a 1965 psychology text with an inquiry that is notable for its carefulness, explicitness, and thoughtfulness.[18] Though he uses the vocabulary of "inherited bodily changes" (mutations) and the like, he is cognizant of attacks on the learned-innate distinction. He defends it because experimental control and variation can be used "to discover which reactions occur primarily as a function of inherent factors and which occur primarily as a function of environmental conditions such as training and imitation" (1965, p. 49; all references are to this volume). It is unnecessary to criticize this statement here, but clearly Munn, unlike many more recent writers, does not claim to reject the nature-nurture distinction, though he is determined to treat it with as much rigor as possible.

In describing unlearned responses as relatively stereotyped (presumably across individuals, not just within individuals) and as appearing without training, he makes the familiar conflation of learned-unlearned with variable-constant; to explain how responses may be unlearned without being present at birth, he uses the concept of maturation. The acquired and the learned are equated and contrasted to the "phylogenetically derived," which must be universally present in normal members of the species (p. 49). But, he continues, not everything that is universal is unlearned, and he gives an exceedingly interesting example, an observation by Fernberger that rats housed in large, badly crowded cages will begin to sleep hanging by their teeth from the wire mesh.

Fernberger (1929) described this as an unlearned behavior appearing only in certain circumstances, but Munn advocates considering it an "individually" acquired habit (notice the assumption that unlearned behavior is not individually acquired), appearing uniformly because of common motivation, structure, and environmental influence. Munn does this not

because he has a precise idea of learning processes that allows him to clas-
sify observations accordingly (if he had relied on his stated criteria for
learning—training or imitation—he would not have been able to dub this
wire hanging "learned"), but rather because his conceptual system does
not comfortably allow it. He cannot imagine the rat's "phylogenetically
derived . . . innate behavioral heritage" determining exactly this behavior
(p. 50). In other words, Munn appears to regard unlearned behavior primarily as an indicator of the rat's most fundamental nature, and since wire
cages seem alien to that nature, leaping and hanging from them must be,
too, even though Munn clearly believes that *any rat would spontaneously
behave in this way under these circumstances,* without needing to imitate
and without needing to be taught. He even coins a term, "coenotrope,"
or "universal habit," to express this belief.

It is instructive to look again at Munn's description of this putative coenotrope. It is acquired because of common motivation, structure, and environmental influence, because it is the only, or most adequate, response
given the circumstances. (It is not clear to me that this response is in any
absolute sense the only or most adequate one in this situation, but it may
be the only or most adequate one *given rat characteristics.* This is exactly
the point.) How are we to distinguish these criteria from the similar genes
and early conditions of development, and therefore similar motivation and
structure, that define unlearned behavior (p. 49)? The only guide Munn
gives is his own conviction that the rats' genes could not determine this
behavior the way they determine innate behavior. The fact that wire hanging appears only under certain conditions (in this case crowding in large,
hot cages) does not distinguish it from unlearned behavior, for many behaviors that Munn would surely accept as phylogenetically derived appear
only under certain circumstances. In fact, the requirements for species-
typical behavior are not limited to the similar genes and *early* conditions
that Munn specifies, but extend to the entire species-typical life cycle.

If Munn had been willing to relinquish the vision of a single set of unlearned responses, he would have been able to meet the hanging rats on
their own terms, applying explicit criteria for learning and pronouncing
the behavior learned or not according to the outcome of inquiry. The problem is, of course, that he lacked, and we still lack, adequate means for such
inquiry, and worse, that the assembly of such means has been *retarded* by
our insistence on defining learning negatively by contrasting it with the

species-typical, the heritable, the developmentally stable, the physical, the congenital, or the maturational.

Frank Beach voiced similar complaints in 1955. Since then, there has been wider acceptance of the notion of species-typical learning, but as we have seen in Fishbein's work, for instance, there remains a belief that some of this is biological and guided from within (necessarily typical), while some is environmental and imposed from without (accidentally typical?).

When anthropologist Marvin Harris, known for taking the "culture" side in a running debate with sociobiologists, states that we still cannot distinguish "with certainty" between genetically and culturally deter-mined universals among humans (1983, p. 21), he is implying a distinction parallel to the one Munn makes between unlearned behavior and universal habits, and also to the one between canalized and uncanalized but univer-sal learned behavior. One wishes for a specification of the kinds of things he, or anyone else, would consider data on this issue. What would a choice bit of evidence look like? Since both kinds of universals are by definition characteristic of the species, literal species-typicality obviously cannot be a criterion. (I use "species-typical" descriptively.) The notion of uni-versality, in fact, is itself equivocal. In Munn's terms, "universal" clearly doesn't mean "rats everywhere and always do this," but rather, "any rat would do this under these conditions." But behavioral theorists tell us that any rat will also push a lever for its lunch under the right conditions, and ethologists say that a rat will build a nest or care normally for its young only under the right conditions. A plant may take one form under certain conditions and a different form under others. An ant larva may become a worker or a queen depending on nutrition, temperature, and other vari-ables. What, then, is universal in each of these cases? A universal potential or ability would bring us full circle to the biological capabilities that sup-port unlearned and learned behavior alike, and that underlie both constant and variable phenotypic features.

The point is not that we should ignore the differences between nor-mal and abnormal environments, which is probably the important thing about Munn's example (a fact that is obscured by his focus on the sup-posed developmental mechanisms underlying the behavior). Nor need we overlook the difference between universal and variable features of normal ones. The point is that these do not distinguish between unlearned and learned behavior or between kinds of developmental processes. Furthermore, when one pursues these questions, it becomes clear that workers who seek to make such distinctions are often doing so not so much because they are interested in learning processes per se as because they are seeking to define nature by the method of exclusion.[19]

Knowing that a given behavior is or is not learned in a particular way may tell us something about ourselves; assuming that we have firm convictions about the difficulty of altering the conditions under which it is learned, it may even tell us how likely it is that the behavior will be learned in the future. (We should not, however, assume that it is only the *altering* of conditions that takes effort and knowledge; sometimes the conditions for learning change despite our best efforts to *maintain* them. Furthermore, much is learned that is not taught, and the subtlety of learning has not, in my opinion, yet been captured by our theories.) We may indeed find out something about ourselves as a species, our possibilities and limitations, by reading the archaeological, ethnographic, ethological, psychological, and anthropological universes, but we need to be clear about what constitutes a message, and what such messages can signify.

c. The basic as special personality traits. The final example of conceptual triangulation concerns not what is basic about some kinds of species-typical development or about certain parts of a species's behavioral repertoire, but what is basic to individuality. Plomin and Rowe (1978, p. 280) describe their search for the "temperamental core of personality," which they define as "those personality traits with an inherited component." When they admit that inheritance is an unsatisfactory criterion because many traits show "heritable influences," they understate the problem. Different temperamental traits will show "heritable components" in different studies, because, as has often been pointed out, heritability is not characteristic of traits at all, but of variation in particular settings and in particular populations at particular times. Nevertheless, they do not abandon heritability but imply that it is a necessary but not sufficient condition; more indicators are needed to "whittle down personality to its temperamental core." The next candidate is "adaptiveness and the phyletic history of temperaments," but the authors here express reservations on the grounds that this is speculative; there is also a tendency for directional selection to reduce variability, they say, and therefore to lower heritability. They thus indicate that they maintain heritability as a prime consideration, one by which others may be evaluated. They still feel that it is useful to look for "directional adaptiveness," to assume that traits are adaptive at moderate levels, and to look for similar traits in other species, especially close phyletic relatives (pp. 280–281). Additional criteria for temperament include early appearance and longitudinal stability,[20] which are considered to characterize "evolutionarily significant" behavior, even though, the authors concede, "adaptive, genetically influenced" traits are not necessarily present early in life or developmentally stable.

So it goes in the conceptual labyrinth. Adaptiveness is treated as an indicator of temperament. Early appearance and stability, in turn, indicate adaptiveness. Heritability is treated as basic to the definition of temperament even though it admittedly specifies very little, is erratically related to the other indices, and is not properly applied to traits in the first place. According to the authors, the temperaments that survive this conceptual "sieve" are emotionality, activity, sociability, and impulsivity (p. 282). Now it may be the case that these are significant, consistent, and methodologically productive dimensions on which to range human psychic variation. That is hardly the point, though if they were not, one assumes that the authors would not bother. The point is that the empirical or theoretical importance of these dimensions, their human and scientific meaning, their degree of centrality in the organization of individual psyches, does not derive from any particular characterization as biological or inherited; it must be demonstrated in the economy of human lives, not assumed. If such meaning can be demonstrated, it is unnecessary to search for additional justification, especially if the justification is as ragged as that described here. If it cannot be so demonstrated, no number of heritable components will compensate.

In trying to tie temperament to some notion of the inherited, Plomin and Rowe not only add confusion to their treatment without increasing its substance; they also obscure the significance of their work. By treating stability and adaptiveness, for example, as indicators of temperament, albeit imperfect ones, they suggest that stability and adaptiveness are not themselves the focus of interest. Since the definition of temperament lacks coherence, however, it is not clear precisely what *is* the point of the inquiry.[21]

The entire network of assumptions and associations made by these authors is nevertheless supported by the tradition of viewing the hereditary, however defined, as basic. It is, of course, just the notion of "basic" that requires definition. Is it a set of factors that accounts for most of the variability of human behavior in a wide range of situations? Is it a set of characteristics that, regardless of its usefulness in accounting for variability, seems to provide organization and coherence to certain personalities? Is it any characteristic that appears early and remains in evidence throughout life? Other criteria are possible, of course, and whether they are related to each other is an empirical question whose connection to heritable variance is dubious. To assume that the presence of a trait in other primates increases the probability that it will be important in the organization of human personalities is questionable at best. This search for homologues in phyletic relatives is ironic in light of another tradition, more venerable than the one that says the inherited is basic. This older tradition treats as fundamental to humans and their diversity just those traits that are not found in other species.

D. The Ghosts in the Ghost-in-the-Machine Machine

The search goes on for the chimerical genetic essences underlying individual or species characteristics. We doggedly stalk through the phylogenetic underbrush and the thickets of heritability coefficients, pursuing the hidden reality that will unify and categorize the variety of the living world. Whether the spectral essence is sought in the form of programmed development toward genetic templates, universal repertoires of unlearned behavior, or inherited core temperaments, the form of the questions we put to nature becomes numbingly familiar, in spite of increasingly impressive jargon and obligatory disclaimers. Given my insistence that such quests for single hidden realities are futile, it would be ironic if I were to posit a single explanation for the constancy, historical change, and variety exhibited by the nature-nurture complex. It has multiple sources, and it shows organism-like vitality, flexibility, and conservation of some of its aspects. But ideas are not things, and they do not act. They are our ways of thinking, and, as has been suggested in these pages, they derive from certain habits of organizing our internal and external worlds, as well as from particularly powerful assumptions and images, traditions of research and interpretation, conceptual confusions, and peculiarities of the history of

philosophy and science. Thorough, authoritative treatment of these issues
is beyond the scope of both this book and my abilities. I can do no more
than suggest some possibilities.

Our tendency to conceptualize our own mental activities in terms of
subject-object relations and of the inner-outer dimension has been noted,
as has the evident ease with which we project these notions inward and
outward to explain all manner of creation and change, stability and in-
tractability. We have an ancient heritage of thought about essence and
appearance, form and matter, about the necessary as universal and the
contingent as variable. These ideas are so interwoven and so deeply entrenched in our intellectual tradition that it is difficult to think in other
terms. Attempts to find alternative ways end up being complicated and
obscure, thus only contributing to the conceptual inertia they are challenging. Certainly Cartesian machine-bodies, as "*instruments* by which
actions initiated from outside could be transmitted or modified" (Toulmin, 1967, p. 823), provided a ready framework for genes as builders of
the machines and as their inner operators as well. The grounding of modern social science in the disputations of the rationalists and the empiricists
not only provided scholars who followed with the will to ask certain questions about mind and society, but also shackled them to certain ways of
construing the possible role of experience in the formation of minds and
the relation of individuals to their societies. Minds and individuals are
regarded as having an inherent universal nature prior to, separate from,
and sometimes in conflict with what is acquired from (or imposed by) the
sensory and social worlds.

The historical fact that biologists study bodies as well as the natural behavior of nonhuman animals, while psychologists, sociologists, and cultural anthropologists study human behavior, made it quite predictable,
though not logically necessary, that whatever human behavior could be
likened to animal behavior (because it seemed automatic and irrational,
because it resembled behavior found in other species, etc.) would be considered biological. Meanwhile, asking other animals to do what humans
were thought to excel at, namely, learning, became psychology, not biology. We have, for whatever reasons, a peculiar relationship between
the behavioral and the biological sciences, a relationship in which some
portions of the "higher levels" are considered really the province of the
lower ones. Some behavior, feelings, or institutions are "genetically de-

termined," and therefore biological, while the rest are the proper material for the behavioral scientists. It is as though a chemist were to say that some compounds were really physical while others were (merely) chemical, or a physiologist, that some biochemical processes were chemical and others only physiological.[22]

Those who study humans, then, have constantly had to choose between regarding that "biological component" of their subject matter as minuscule and unimportant and seeing it as large and significant, and that is what much of the nature-nurture debate has been about. The association of the biological with the immutable and the psychological with the malleable ensured that biologists and social scientists would tend to line up on opposite sides of arguments about the possibility of change, though just because such practical conclusions are not properly drawn from the premises, and because the relations among these ideas are so unkempt, this self-assembly is not unerring. Until relatively recently, we confined ourselves to attempts to delineate the boundaries between these two domains, often declaring them to be fuzzy indeed but rarely doubting the existence of different territories. We have tended not to question the assumption that nature is defined at a level other than that of the individually lived life, regardless of the circumstances of that life. These ruder questions have yet to make serious inroads into scholarly or popular thought.

Though Konner and others are correct in pointing out that any particular view of this inherent nature is compatible with a variety of political systems (recall, however, that he implies that a biological view is necessarily a dark one), and though it is therefore not possible to derive a particular moral or political system from biology, there seems to have been some tendency for empiricism to be turned to the purposes of liberals, leaving the "biological" to be exploited by conservatives. Despite occasional raids across these broad but tenuous frontiers by progressives who find an ethological or paleoanthropological justification for their claims, or by conservatives who point out that poverty is just reinforced by welfare payments, I suspect that even these diffuse and shifting political and moral alliances have an appreciable impact on the way we think about development and the nature it produces.[23]

This phrase, *the nature it produces,* is, of course, carefully chosen. Nature is here viewed as a result, not an initial condition; multiple, not unitary; and inseparable from particular developmental circumstances. This does not make humans or any other creatures "infinitely malleable," either in the sense that they are or can be infinitely variable (or absolutely uniform) in all their aspects, or in the sense that characters can always be altered in an individual's lifetime. Though constructivist interactionism has sometimes been seen as threatening to make our view of development overly complex and not amenable to generalization, perhaps we are preventing ourselves from discovering new kinds of generalizations by insisting on the old ones (P. P. G. Bateson, 1983). The study of persons-situations interaction, for example, is by no means dependent on conceptions of personality as inherited or as unresponsive, and in fact is limited by them. Yet it offers the possibility of finer-grained analysis than either traditional personality or situational approaches did (see Chapter 2). Similarly, ethological studies are increasingly describing kinds of organism-environment relationships rather than cataloguing innate patterns of behavior.

Another factor that may help perpetuate nature-nurture ideas is related to those mentioned above but deserves special mention. This is what we might call their mythological function. Toulmin observes that a scientific myth must be understood not in a strictly scientific sense, but in an extended one, involving other kinds of motives (1982, p. 62). Some of the preceding discussion has been an attempt to identify some of those extrascientific motives for distinguishing an underlying nature from accretions by experience. Yet another such motive is suggested by Shklar's concept of the subversive genealogy, which documents unsavory origins of an existing system in order to undermine it. The contamination of the present by the past is employed as a rhetorical device.

She observes, as others have done before her, that the need for myth arises when people are "faced with extreme moral perplexity, when they reach the limits of their analytical capacities and powers of endurance" (Shklar, 1971, p. 137). The decline of the influence of the concept of original sin, for example, left us with some uncomfortable questions: If we are inherently good, why is the world so full of injustice and tragedy? Myths help us think about these contradictions and their causes. The role of the Original Man or the State of Nature, when viewed from this perspective, is to help us understand, interpret, and judge the present. Though Shklar's treatment is restricted to the subversive use of such inquiries into the past (which, it is important to note, need not, and often do not, claim historical accuracy), it is certainly possible for a genealogy to be justificatory as well, or to serve some other instructive or inspirational purpose. Whether one sees society as a perversion and suppression of what is natural; as a necessary restraint, perhaps accomplished at considerable psychic cost; as a means of securing agreed upon aims; or as a liberating and elevating release from animal nature, the mythical *Then* is used to characterize the *Now,* often by contrast. This is why emphasis on any particular aspect of biological nature does not specify a single political or moral position. This inferential slippage, in fact, might alert us to the possibility that something other than straight empirical reasoning is being employed.

Cartmill (1983, pp. 77–78) comments that myths, whether they are factually true or not, express deep feelings and ideals, and that "whatever such stories single out as important factors in our origins become important parts of our self-image. Conversely, such stories won't be listened to unless they account for the human peculiarities that we think are crucial markers of humanity." He describes the implications of the current image of Man, the Killer Ape, for our view of ourselves as "antagonists of life" and points out that it may be regarded not as an excuse for war and violence, but as a "cautionary myth" that may move us to guard against our own death-dealing capacities. Even though the significance of findings on our species's origins for social policy or even for scientific prediction is negligible, this paleoanthropologist feels, the ideological function of such myths deserves attention. The Killer Ape, furthermore, is not the only image we see when we peer into the evolutionary looking glass. Where some of us see the monsters we fear we really are, others glimpse the prototypes for the world we would like to build, and still others behold the peaceful paradise we fear we have lost forever.

This way of looking at conceptions of nature helps make intelligible the degree of emotion and conviction with which people report or deny sexually dimorphic behavior in baboons, for instance, or "murder," "political alliances," or "comradeship" among chimpanzees. It also sheds light on the special meanings certain groups of "primitive" peoples, the past manifested in the present, have had for us—the Native American, first as bloodthirsty savage subdued by European civilization, and more lately as paragon of ecological wisdom; or the hunter-gatherer, whose bare, stressful struggles for subsistence are contrasted to our own advanced society. (Netting, 1977, pp. 9–10, points out that more recent research has shown hunter-gatherers' lives to be full of leisure and not nearly as precarious or desperate as we were pleased to believe.)

The conditional nature of "nature" being advanced here does not impoverish the concept, though this is what is implied by alarmed cries of "But that would make mother love merely *learned,*" or by satisfied exclamations of "So obviously sex roles are nothing but cultural conditioning." Rather, nature takes on substance in the real world, richer and fuller than the phantom outlines, constraints, potentials, or norms of reaction it is normally granted.

There are no ghosts in machines,[24] only persons in the world, thinking, feeling, intuiting and sensing, deciding, acting, and creating. And there are therefore no ghosts in these ghosts, no programs in the operators of the machines, making them feel as their ancestors felt, making them act or want to act as gorillas or chimps act. But there *are* many ghosts in the psychological, social, and cultural machines that create and re-create the body-machine, the ghost in it, and the ghost in *it.*

7 The Ontogeny of Information

The ways of thinking we have examined, then, seem to persist partly out of historical inertia and partly because we have concerns that do not dissipate when technical refinements are made in the conceptual apparatus we have always used to address them. The remedy is surely not a draconian prohibition on simile, hyperbole, and the poetic imagination. Literary license, though, is not license to mislead, and when metaphor is employed in the service of scientific understanding, it should be accurate and helpful as well as vivid and evocative. It is obvious that I do not believe it necessary to continue down the same paths we have trodden for so long, and the alternative is distributed holographically throughout the preceding chapters on vital form, change, and variability; on biological metaphors; and on the gaggle of ghosts that seem to guide the reasoning of so many thinkers about our place in the living world. I hope that this other approach has become familiar to the reader as we have examined these problems. It is likely to resemble a truism when stated briefly and explicitly (which is why various "interactionist" credos have become so empty) and gains utility and meaning only if consistently applied in thinking about a great number of particular issues. Nevertheless, it may be useful at this point to try to articulate it in a more unified manner. What would it mean to synthesize the battery of dualities that has populated the intellectual landscape for so long? It is clearly not enough to reject one and choose its successor in the next breath. We need a way of addressing both the important scientific issues (such as the control of biological processes and the significance of variability and constancy) and the more philosophical ones that we encountered in the preceding chapter, a way that describes the world of investigated phenomena more fully and consistently than the old dichotomies can and that helps us see whether, when, and how those observations are relevant to more general questions of inevitability, desirability, and reality. We will look first at the issue of control of development, then at developmental systems as networks of mutually dependent controls and effects evolving over time.

A. Emergent Controls

Control of development and of behavior may be said to emerge in at least three senses. First, it emerges in interaction, defined by the mutual selectivity of interactants. Second, it emerges through hierarchical levels, in the sense that entities or processes at one level interact to give rise to the entities or processes at the next, while upper-level processes can in turn be reflected in lower-level ones. Third, control emerges through time, sometimes being transferred from one process to another. These phenomena are not independent; in fact, they are three ways of looking at constructivist interaction.

I. DEFINITION OF CONTROL IN INTERACTION

Placing determination of form in the genes requires a choice: either (1) some phenotypes are formed (mostly) by the genes and some are formed (mostly) in some other way (the conventional formulation), or (2) all possible outcomes are provided for by the genes and simply selected by circumstance. The first choice, as we have seen, is unworkable because there is no satisfactory way of separating genetic outcomes from environmental ones. This is not because we lack the adequate techniques, or because, as nearly everyone agrees, "The extremes are nonsense: it is the *degree* of genetic or environmental influence, control, or constraint that counts." Degrees or relative amounts are nonsense for the same reasons dichotomization is. The choice is unworkable because no criterion allows the specification of one kind of control or constraint without specification of the other. What an instance of genetic transcription or a sensory stimulus controls, if it controls anything, depends on the context. The common tactic of taking large assemblies of factors for granted ensures confusion when there is disagreement over what is or should be included in those assemblies and establishes an ontological double standard whereby some developmental phenomena are more basic, real, or biological than others.

Though the contingency of the effects of both genes and environment tends to be dismissed with an impatient "Yes, yes, of course" by workers in the field, the full implication of such interdependence seems not to be so clear, since genetic constraints and rules, biologically based traits, and the rest of the nature-nurture menagerie still appear in the discussions of the most dedicated "interactionists," who themselves devote ample space to denouncing the nature-nurture distinction. But a constraint (one does not hear of environmental constraints, evidence that the interaction argument has not extended this far) is no more autonomous than a factor, information, or an influence.

The second choice, placing "determination" of all potential in the genes, avoids these difficulties by being vacuous. The genome is capable of only those interactions of which it is capable. If the full, all-possible-contingencies plan is in the genome, and if, in general, the genome remains the same through ontogeny, then actual specification of developmental events must be "nongenetic" as well as genetic: by developmental state (because the cell, tissue, or organism responds differently to given conditions at different times) and by surrounding conditions (because a given biological entity in a given state responds differentially to various conditions). Note, however, that this is not an "environmental" explanation. When a stretch of DNA is transcribed, that event, whose determination must itself eventually be explained, might determine some biologically important fact, and the results of past transcriptions and translations may well constrain processes in that cell or in others. Far from being slighted or taken for granted, the genes in this view are given causal efficacy in a way denied them in more metaphorical accounts simply because, apart from the obvious difficulties with the idea of completely autonomous genes spontaneously arranging their own activity, unresponsive control cannot be effective control. Whether one is speaking of machines, organisms, or human affairs, control without feedback tends to become derailed and therefore useless or destructive. Although it is commonplace to use the language of cybernetics in biology and the social sciences, the fact is that cybernetics is inconsistent with the equally common attribution of true causal power to one entity in an interaction.[1] Feedback has been described as control on the basis of actual, not expected, performance (Wiener, 1967, p. 36). Its very definition is the ability to control by being controlled. It is, incidentally, the influence of results on the processes that produce them that I take to be central to feedback. The implication of an explicit set point, the mechanical counterpart of an expectation, while useful for understanding servomechanisms and, perhaps, for certain simulations, can be misleading in treatments of biological processes; it lends itself too easily to naive notions of static genetic control. (More on changing controls later.) People who, by "programmed," mean inevitable and caused only by the genes are forgetting that a computer program does not guarantee an output regardless of input. On the contrary, the more sophisticated the program, the more subtly it responds to its input. A program whose output were completely specified by ("completely controlled by") the program itself would be of limited value.

Every developmental interaction, and therefore the entire norm of reaction, is jointly determined. To question the notion of genetic constraint or potential is not to deny limits to variability, flexibility, or adaptation. It is simply to give all developmental interactants comparable theoretical status, to recognize the importance of levels of structure above (and below) that of the genes, to insist that the significance of an interactant must be discovered by investigating the roles it plays in ongoing processes, and to point out that phenotypic potential cannot be said to be limited in any practical sense until we know what the limits of genotypic and environmental variation *and interactions* are. Though we may have little control over the first term in this statistical trio, the ranges of the second two, and therefore the effective range of the first, are indeterminate and to some extent, perhaps small, subject to our intelligence.

Effective control, then, depends on the specifics of an interaction, its context, and on the history of the system in question.

2. HIERARCHICAL LEVELS OF CONTROL

One scientist's given is often another's analytic goal. Those who study embryogeny must account for just the anatomy and physiology that others take for granted as genetically programmed. Whether or not they accept programming as causal explanation, it is *particular* workings that must be specified; thus cognitive metaphors cannot substitute for analysis and careful model building. What one finds when one looks at morphogenetic models, for example, is an assumption, if not an explicit declaration, that structural givens have a history, that control of the present is itself developmentally contingent, not only on prior interactions, but on interactions at lower and higher levels. Descriptions thus involve multiple control of processes, shifting with the operation of those processes as physical structures and biological products are built up, deformed, broken down, distributed, or destroyed. Organs are formed by tissue changes, and organ functioning feeds back to influence tissue growth and maintenance. Computer simulations are central to some models, but it is important to note that no particular epistemological respectability is guaranteed, as Miller et al. (1960) indicated, by placing a switchboard or punched card in the head or cell rather than a clerk or operator. If there were, our problems would have been solved long ago by replacing the man in the head, or in the cells that constitute the head, by a recording of his voice. Instead, the value of the model is finally judged by its ability to fit and predict observations, and computers may be used to test this ability. People who work at this level, that is, gain little by saying that a cell that moves in a certain way is programmed to do so; indeed, if a program metaphor is used at all, its hows and whys must ultimately be expressed in terms of units and events in the real world—electrical charges, chemical gradients, mechanical pressures, enzymatic reactions, and so forth.

Elsdale (1972) describes certain "inherently precise" stochastic processes that produce order from basically random input. Rather than being channels through which form is transmitted from a shaping agent to a product, as in manufacture by lathe, in which variation is an inevitable intrusion to be minimized, an inherently precise process utilizes simple, imprecise operations to produce exact results. An example is the shaping of a spherical lens by means of small, loose grinding motions between two roughly shaped blocks, one slightly concave and the other slightly convex. Many random interactions between the blocks result in an increasingly perfect lens surface. Unpredictable variation, far from being noise, destructive of a pattern being transmitted from source to product, is itself the source of pattern. But notice that the rounded form exists neither in the rough blocks, nor in the loose movements of the joints on which they are mounted, nor in a plan or program separate from the operation of the few simple parts. It arises from "random energetic inputs and mutual constraints by the parts" (p. 102). It might be argued that the form does exist in the mind of the designer of the machine, but this is built, so to speak, into any machine example; and while the intention of the designer in some sense explains why *this* device is employed to make a lens, it does not explain the mechanical workings themselves. In any case, the principle does not require an engineered machine. A small pebble in a cavity in a larger stone may be tumbled by tides to produce a smooth, regular bowl.

Elsdale (1972, p. 102) then describes morphogenetic movements of cultured fibroblast cells, applying the inherently precise concept to the emergence from random cell movement of orderly three-dimensional arrays of layers of parallel cells, each layer perpendicular to adjacent ones. What is crucial about the cell movements is that they are relatively small in magnitude, spontaneous, and individual, and not organized at the aggregate level. Their interaction produces constraints on further movement such that an organized array results. Though they result from random movements, the arrays of cell sheets at right angles to each other are not themselves "chance" phenomena. Nor are they necessary, in the sense of being fated by a plan in the cell or in some organizer, directing each cell to a particular position or even in a particular direction.[2] They become necessary, however, given an assemblage of particular kinds of cells under particular conditions. If the enzyme collagenase is added, accumulation of collagen is prevented and movement of cells over each other is inhibited. Sheets eventually form, but no three-dimensional layering occurs. Though collagenase in this case is an "artificial" addition, destruction or alteration of biological materials is part of many developmental processes; the point is not that any of the necessary conditions in Elsdale's preparation could be altered experimentally, but that conditions in ontogeny actually *do* change, and that these sequential, interdependent changes are the essence of development.[3]

Once cell layers are formed, they can enter into higher-level morphogenetic interactions, which in turn may bring about further changes in cell activity. Cells both constitute and are controlled by the tissues they construct.

Though the inherently precise process is only one approach to morphogenesis, there are others, some associated with computer simulations, others not. (See Goodwin and Trainor, 1983; MacWilliams and Bonner, 1979; Newman, 1984; Ransom, 1981; papers in Waddington, 1972; Weiss, 1966, 1967, 1978.) Causality in these treatments is mutual, mobile, complex, and dependent on phenotypic state as well as on the surroundings. Control has an ontogenetic history, and both controls and constraints must be understood hierarchically.

Brian Goodwin, whose genetic hypothesis concept was criticized in Chapter 5 but who was also an early critic of the computer program metaphor (1970), has continued in the direction that was already evident in those writings in the early 1970s. That direction, toward a fuller and more sophisticated appreciation of developmental processes as absolutely central in generating and constraining evolutionary possibilities, is being more and more clearly defined by a group of thinkers who are challenging traditional views of development and evolution in critiques that often show considerable overlap with the arguments in the preceding pages. (See Goodwin's 1982 comments on internal and external stimuli and the inability of genetic programs to explain the phenomenon of the phenocopy; see also Alberch, 1982; Ho and Saunders, 1979, 1982; Løvtrop, 1981; Webster and Goodwin, 1982.) Though I am not in agreement with the emphasis in certain of these writings on necessary and universal forms, I view these efforts by developmental biologists as important to the present project, for they are redefining embryogeny in ways that help to eliminate the opposition of nature to nurture that has plagued both scientists and laymen while hindering rapprochement between evolutionary and developmental studies. As long as genetic determinism is accepted for morphology, students in the life sciences will continue to look for thoughts, feelings, and behavior that seem to fit that model and to contrast them with others that must then be seen as developing according to different principles. And as long as evolution is thought to require unidirectional genetic control of the development of inherited form and function, full integration of studies of ontogeny with those of phylogeny is unlikely.

3. TRANSFORMATION AND TRANSFER OF CONTROL OVER TIME

The only way out of the problem of predetermined potential, which is an attenuated version of predetermined forms, is thus to see potential itself, in the sense of possibilities for further alterations in a given structure, as having a developmental history. It is multiply, progressively determined, with new varieties of causes and consequences emerging at different hierarchical levels and with time. Control, which is a kind of relationship, is defined in interaction in the manner described above. But a more complex, higher-level dependence of control on process must be recognized as well. In describing types of feedback, Wiener observes that if information fed back to the system changes the system's "general method and pattern of performance," learning may be considered to have occurred (1967, pp. 84–87). It is not accidental that this description fits development in general, since I think the traditional ways of distinguishing learning from other kinds of development are often difficult to defend. In any case, what is important here is that the machine's "taping" (programming) is itself altered, and with it the manner of processing data.

In a paper on developmental rules, P. P. G. Bateson (1976) describes the "interaction" and "control" approaches to development (roughly equivalent, though he does not use the terms, to epigenetic and preformationist emphases, respectively) and suggests that integration of these traditions requires considerably more than "a trivial compromise . . . whereby it is admitted that some patterns of behaviour develop totally as a consequence of interaction between the animal and its environment and others are the products of predetermined, self-correcting developmental processes." What should be investigated is the ways in which "many developmental control mechanisms are themselves modifiable by the environmental conditions in which the animal grows up" (pp. 410–411). His discussion includes alteration of the "preferred setting" in avian imprinting, in developmental regulation of weight (with stunting as resetting of the preferred weight), and in social interaction. In the course of these treatments he mentions the shifting from one control mechanism to another upon satisfaction of particular developmental criteria and the importance of focusing on the environmental conditions and developmental states associated with such transitions.

In stunting, the alteration of the preferred setting is a source of variation among individuals, and Bateson mentions the possible adaptive consequences of such variation. Even in the species-typical case, however, the setting is progressively altered to allow growth. In Bateson's model, stunting results from making the size of the increment in preferred weight depend on the amount of discrepancy between preferred and actual weight. The incremental changes in the set point are basic to the model. Even though, as he points out, the realities of research often focus attention on variation in conditions that result in differences among individuals, and even though this is what is generally meant when an investigator says that a phenomenon is modifiable by environmental conditions, control mechanisms must also be modified in the species-typical development;

development *depends* on these modifications. Due to the frequent confusion between changes in individuals and variation among individuals, discussed earlier, *alteration, modification, change, flexibility, malleability,* and the like tend to be associated with departures from the species type, while the usual developmental pathway, maturation, is attributed to rigid genetic control. But, as we have seen, saying that the genome controls maturation does not illuminate the system of shifting controls, the sequence of modifications, that actually constitute *any* development, typical or not. (See also Hailman, 1982, on ontogeny as changes in control; and Pribram, 1980, on the TOTE as a homeostatic model and on the importance of alteration of set points in such models.)

Controls, preferred settings, constraints, programs, rules, plans, and information, then, all have contingent developmental histories.[4] The cognitive-causal models we have considered have tended to present these controls as ahistorical (or rather, as having a phylogenetic history but not an ontogenetic one), which provided no satisfactory way of explaining either species-typical development (maturation) or the manifest flexibility and multiplicity of many developmental phenomena, save the declaration that all contingencies were somehow anticipated, or at least hypothesized, by the DNA.

Fate is constructed, amended, and reconstructed, partly by the emerging organism itself. It is known to no one, not even the genes. Lewontin (1983, p. 27) points out that Platonic ideals survive in biology, especially among developmental biologists, who are "so fascinated with how an egg turns into a chicken that they have ignored the critical fact that every egg turns into a different chicken and that each chicken's right side is different in an unpredictable way from its left."

Saying that controls and constraints emerge in developmental systems (phenotypes and their surroundings), and that they may emerge in different ways in different lives, is not a mere quibble. It is the only formulation that is consistent with the observations of developmental, genetic, and comparative studies alike, that allows discussion of the typical and the atypical, the stable and labile, the intra- and extraorganismic, without the entanglements of a vocabulary that is vague or circular at best and contradictory at worst. It does not, however, allow pronouncements on the normal or the desirable, apart from observations of particular patterns of incidence and consequence, respectively, or outside a particular framework of values, and the inevitable cannot be characterized separately from its context. The real trick in prediction is knowing how much of the context must be taken into account, and only knowledge of the constructive process will give insight into this question. Increased understanding of embryonic development has diminished our concern over whether a pregnant woman sees particular animals that might influence the shape of her baby, for example, and has heightened our vigilance over the chemicals she ingests or the stress she undergoes.

What all this means is not that genes and environment are necessary for all characteristics, inherited or acquired (the usual enlightened position), but that there is no intelligible distinction between inherited (biological, genetically based) and acquired (environmentally mediated) characteristics. Heritable and other kinds of variation can still be calculated in particular investigations and may direct inquiry in useful ways or indicate the possibility of selective breeding in a population, but such statistics do not allow partitioning of phenotypes.

Once the distinction between the inherited and the acquired has been eliminated, not only as extremes but even as a continuum, evolution cannot be said to depend on the distinction. What is required for evolutionary change is not genetically encoded as opposed to acquired traits, but functioning developmental systems: ecologically embedded genomes. Whole ensembles of developmental interactants and processes, temporally and functionally integrated, evolve with time. Exclusive focus on genes has virtually guaranteed that anything with an evolutionary history would be seen as being created by those genes, often in a peculiarly atomistic "gene for character" manner. But if we detach ourselves from the notion that characters can be transmitted, we can see that what is required for evolution is that similar sorts of processes eventuate in similar sorts of products, again and again, and that variation in those processes and products have some impact on subsequent constructions. One of the three basic ideas of evolutionary theory as described by Lewontin (1978) is the principle of heredity, defined as closer resemblance between relatives than among unrelated individuals. (The other two are variability and differential reproductive success.) Such resemblance, however, can be due to any number of factors, some of which may be quite stable (lineages may be nonrandomly distributed across habitats, for example). The developmental system, then, encompasses not just genomes with cellular structures and processes, but intra- and interorganismic relations, including relations with members of other species and interactions with the inanimate surround as well. This is all implicit in present evolutionary theory, and occasionally explicit, just as the principles of reciprocal selectivity and the development of control and constraints are implicit in the observations of embryologists, molecular biologists, and the rest of the life sciences. What I am offering here is not some astounding new fact that requires discarding large chunks of theory. Rather it is (1) a plea that we say clearly what we, in some sense, know; (2) a way of doing so; (3) an articulation of the full implications of doing so; (4) some speculations on why it has been so hard for us to do so; and (5) some reasons it is imperative that we do so.

The ways in which controls and constraints emerge through reciprocal selectivity in interaction, through levels and through time, offer a way of looking both at ontogeny and at phylogeny and, what is perhaps more important at this time of increasing dialogue between workers in these two areas, a way of relating them.

B. Developmental Systems in Ontogeny and Phylogeny

I. SELECTION AND DEVELOPMENTAL SYSTEMS

What shapes species-typical characters is not formative powers but a developmental system, much of which is bequeathed to offspring by parents and/or arranged by the developing organism itself. The same is true of atypical ones, which may result from developmental systems that are novel in some respect; an aberrant climate or diet may "play" on the genome in a different way, a mutation may eventuate in altered stimulus preferences or metabolic processes, thus altering the effective environment, etc. This is why it is important to apply the term "ontogeny" to all development, whether common or rare. Granted, some inherited sets of interactants will produce a phenotype whose viability is limited, perhaps to a brief moment. Failure may be as fundamental as the lack of fit between two molecules. Developmental effectiveness is a powerful selective consideration.

Certain interactants are regular features of the environment, narrowly or broadly construed, and do not depend on anything the organism does, though the utilization of these resources is dependent on the character-istics of the organism. The availability and qualities of others do depend to some degree on the organism itself, and the organism's ability to pass them on is quite as crucial to its reproductive success as its ability to pass on its genes. In pursuing the metaphor of the intensely individualistic, selfish gene, manipulating not only the body in which it resides but other objects and bodies as well, Dawkins (1982, p. 236) arrives at a network of influences and constraints quite as widely ramified and intricate as the one being described here, though he is at pains to distance himself from the sort of sentimental glorification of the great web of nature that he as-sociates with "pop ecology" and that he dubs the "BBC Theorem" after the rhetorical style of certain television nature shows. It is, in many ways, the same network, viewed from different sides.

In presenting his gene-centered view of life, Dawkins (1982, chapter 13) refers to the Necker cube. Well loved by students of perception, this is a two-dimensional line figure, a pair of overlapping squares connected obliquely at the corners so as to represent a cube. The figure, however, is ambiguous in the third dimension, seeming first to be a cube with its front face to the lower left, for example, and then a cube with its front face to the upper right. The perceiver can usually flip at will from one percept to the other, though some have initial difficulty seeing both cubes, and therefore cannot flip at all. While Dawkins uses the Necker cube to show how we can see evolution from the gene's viewpoint as well as the organ-ism's. I suggest that we can also use it as a metaphor for a different sort of flip, between seeing causal networks as exploitation and one-way control and seeing them as interdependency and contingency. When he speaks of selection for a gene's ability to get along with (effectively operate in its own behalf by cooperating with) other genes because of shared inter-ests, and when he points out that such "cooperation" takes place not only within a genome but across cell, organism, and even species boundaries, he is acknowledging just this interdependency. I would hardly want to elaborate a metaphor of the trusting, dependent gene, which would only encourage this habit of giving human characteristics to genes and then explaining all manner of bodies and behavior by them, but the potency of DNA is hardly possible without this vast network of interactants and interactions. (See Symons's critique of the selfish gene vocabulary [1979, chapter 2]. He notes that the image of gene-driven robots simultaneously invests genes with motives and divests organisms of them. See also Midgley, 1979.)

Interactants whose continuation is to some degree dependent on their interactions in the ecological web will be differentially passed on (inherited or not by the next generation as parts of the effective developmental system). This is natural selection, which has traditionally been reserved for gene frequency changes. If selection is not redefined to include all developmental interactants, these interactants will still be indispensable to the process, and will have to be assumed, just as they have always been assumed (sometimes at the expense of the clarity of the treatment). The sun, on the other hand, shines whether an organism utilizes its energy or not. It follows, however, from the causal interconnectedness of the real world, that the difference between interactants that are reliably "there," like oxygen and sunlight, and those whose distribution is affected by the organisms that are affected by them, is relative to the size of the investigative framework. All that oxygen, after all, was made by life, and the prospect of a nuclear winter reminds us that life can effectively take the sun away.

It should be clear that "system" is being used in quite a broad manner. Tightly organized, well-regulated physiological functions are certainly included, but to restrict the term to species-typical developmental pathways and structures would defeat the purpose of the conceptual reorganization being attempted here. Sets of interacting influences that may change with time, systems in this larger sense are revealed by causal analysis of more or less organized phenomena. Calling a coral reef an "ecosystem," then, does not imply that it is a large organism or even very much like one. Similarly we may speak of a system of communication (not necessarily codified or planned) in an institution or of an economic system or a kinship system without implying that just this structure has evolved over thousands of years by natural selection. Scientists are usually interested in common and/or enduring interactional networks, but one might want to investigate rare or transient ones as well. A unique historical sequence or an individual life may well be worth our attention.

Since all aspects of the phenotype are products of ontogenesis, they are in some sense acquired. Means (developmental interactants) are inherited, results ("natures") are acquired by construction. A reproductively successful organism passes on the pertinent environment in many ways. This, to a large extent, is what it *means* to be reproductively successful, and it involves much more than having the "right" genes. (The fact that the organism has considerable but not unlimited influence over the set of interactants it will pass on to its offspring would seem to account for some of the difficulty we have in relating the concepts of adaptation and reproductive success.) The cellular, and for some organisms the larger, reproductive system is provided with chemical and mechanical prerequisites for early operation. Nutrition, moisture, temperature, and light are provided by egg structure and placement. Insects that do not give parental care as it is usually construed may nevertheless provision their young by leaving food with the eggs or simply by laying the eggs on or in edible material. Gut symbionts may be passed from parent to offspring and among siblings (Thompson, 1982, p. 77). Margulis (1974) gives an example of hereditary endosymbiosis; the seeds of *Psychotria bacteriophila* contain bacterial symbionts, so that the next generation inherits chromosomes, cytoplasm, and bacteria along with the larger developmental system. The same author cites a protozoan that is a symbiont and that contains three others, a nice example of nested developmental systems. Prenatal stimulation and hormonal influences from mother or siblings and parental care impose more order on an already ordered surround, as do conspecifics, prey, and the organism's own products—odors, constructions, effects on others, and so on. Hofer (1981) reports that the onset of puberty in mice is influenced by the presence of adult males during infancy and the early juvenile period. It would seem, then, that anything that affected the association of males with young would affect puberty onset, whether it had primarily to do with maternal behavior, preferences of the males, group density or social structure, or some other ecological factor. To the extent that such influences are predictably passed on, development proceeds with the automaticity of embryonic development. And to the extent that they are nonrandomly associated with certain lineages, their effects will show heritable variation. The stability of such heritability depends, then, on stability of the developmental systems in question and of their patterns of variation, and this in turn requires not only that genotypes be stable but that relevant aspects of the surroundings be stable as well. The likeness to embryogeny is not mere analogy, and it is not due to special formative forces or to preexisting ends; it is due to more or less close coupling of developmental interactants. (Notice that extremely close coupling need not be the result of the organism's or its parents' actions, but may be a function of general ecological features. Notice also that this view fits well with the notion that an organism may contribute to another's fitness.)

Final or ultimate causes are best understood not as exerting causal influence on the present through the genes but as describing the history of a given disposition of proximate causes. Just because certain kinds of proximate causes are related in certain ways, certain consequences follow, which consequences may enhance the probability that just those proximate causes, and therefore those consequences, will be associated again. (Notice that conventional adaptiveness is not the only conservative influence on the propagation of developmental systems—some argue that it is often but a minor factor—and that the inherent order of morphogenetic and other developmental processes seems to confer much transgenerational stability.) What is important here is the ability of such causal configurations to influence their own conditions, and to do so repeatedly and progressively. An infant rat may need the reassuring smell of its own nesting material to show certain kinds of response alternation, discrimination, and other behavior in learning experiments (Zolman, 1982, citing work by Smith and Spear), but in the normal course of events the nest and its odors come with the developmental territory as surely as the olfactory apparatus that enables the rat pup to sense them. In influencing the pup in this way, the nest may increase the probability that a competent rat will develop, and thus that a new nest and young will interact in the future.

The literature on coevolution (roughly, the reciprocal selective influence of evolving groups) is rich with examples of the mutual dependence of ontogenetic and evolutionary fortunes. Its precise definition and pervasiveness are under dispute (see papers in Nitecki, 1983). The broadest definitions embrace much of evolution, while other workers argue that coevolution is relatively infrequent. At any rate, a bit of reading in this area quickly demonstrates that even if one is not dealing with a coevolved complex, the understanding of a developmental system involves a whole set of interspecific interactions, as well as a host of geographical, climatic, and other ecological factors. Parasitic relationships lend themselves with grisly effectiveness to Dawkins's manipulative gene language (1982), but there are also many mutualisms that involve interlocked benefits. Feeding an insect may be important to a plant's pollen dispersal, or an animal may defend or otherwise care for an organism on which it depends. (Thompson, 1982, in fact, states that mutualisms often evolve from antagonistic relationships.) In all these cases an organism plays a part in another's developmental system, and its survival may depend partly on how it does so. In order to disperse pine seeds, for example, by burying them in caches (thus partially defining the distribution of the tree by selecting cache sites), certain nutcrackers must be able to extract them efficiently from pinecones and distinguish viable seeds from inferior ones (Tomback, 1983). Artifacts may also play such roles (Dawkins, 1982, chapter 11) and may endure longer than their makers.

Though it is customary to speak of morphology as directly inherited and behavior as less directly inherited (if at all), this is presumably a reference to the relative *reliability* of appearance of a skull, for example, as opposed to a particular action, since bones are not, as everyone agrees, literally passed on in germ cells, or anywhere else, any more than breathing and biting are. In this sense, bones can be quite as indirectly "inherited" as movements, though bones are things, whereas movements are what some things do. Even if a metric of directness of influence were to be developed, perhaps by counting metabolic steps, some behavior would surely be more direct than some structures; it would not on that account necessarily exclude experiential efforts or be species-typical or unresponsive to other influences.

In an analysis of causal chains in biology, Dawkins (1982, p. 199) declares that nobody has trouble understanding genetic control of morphological differences (I disagree, unfortunately) and that nobody has trouble understanding that there is no difference in principle between genetic control, in the sense of gene-character correlation, of morphology and of behavior (I disagree again). He argues that "if there is any sense in which the brain is inherited, behaviour may be inherited in exactly the same sense. If we object to calling behaviour inherited, as some do on tenable grounds, then we must, to be consistent, object to calling brains inherited too." This time I agree. One could argue that the point here is subtle, but my position, of course, is that these are very important subtleties. I do indeed object to calling behavior and body "inherited" in the sense of "being made, programmed or transmitted by the genes," and I have argued that one inherits developmental means, not bodies and behavior (or even plans for bodies and behavior). I do not object to the idea that genes can, in specifiable situations, account for differences in either, though I think the word "inherit," which suggests "receive by bequest," is hard to apply to differences, and this makes thinking about already difficult issues even more difficult. Dawkins no longer speaks of genes making the machines in which they go about, though many of his manipulation and control metaphors still invoke the intelligent, deterministic "bogey" he repudiates. Development and evolution *are* linked by the differential passing on (availability to the next generation) of that which is responsible for development; but the genes, as Dawkins understands very well, do not exhaustively define this category.

Some workers speak of "direct" inheritance of cellular patterns and products from previous generations, including initial asymmetry of the egg, which may orient the first division, polarity, and pattern of organelles, certain changes in the cell membrane (Bonner, 1974, pp. 226–227, who speaks, with an interesting combination of terms that are usually opposed, of axis of symmetry being "directly acquired, or inherited from the mother"; Stebbins, 1972), not to mention the organelles themselves, which may be former symbionts, genetically distinct and permanently integrated into cells only after a long history of association with them (see Margulis, 1974, 1981; papers in Fredrick, 1981). "Directness" here presumably refers to the fact that the feature is associated with the germ cell and is thus very reliable.

In an impressive treatment of ontogeny and evolution, Raff and Kaufman (1983, pp. 101–110) illustrate many of these points with a wide range of examples from embryology, molecular biology, and classical genetics. In describing the three information systems in a fertilized egg (nuclear DNA, regionalized cytoplasmic macromolecules, and the "cytoskeletal matrix" or cellular structure), they show that the initial developmental system includes (but, I would argue, is not limited to) the organism's genome, complex cell structures, and messenger ribonucleic acid (mRNA) that derives from the mother's genome, not the embryo's own. In sea urchin embryos, the latter two parts of the system can bring development a surprising distance. Changes in cell shape, assembly of cilia, and synthesis of hatching enzymes will proceed all the way to the blastula stage without any transcription of the embryo's own genes. Again and again these authors cite research showing the regulation of gene activity by temperature, cytoplasmic constitution, and a variety of other factors. The importance of the cytoplasm, for instance, is revealed when nuclei are transplanted into cells from different tissues or even different kinds of organisms. Patterns of gene activity are frequently those characteristic of the *recipient* cell. Surely the contemplation of certain maternal-effect mutations, in which it is not the embryo's own genetic makeup but the makeup of the mother's egg that is defective, calls for a conception of control of development that extends beyond genetic programs and is capable of accommodating other sorts of effects, some of them transgenerational. The unfortunate progeny of the axolotl mutant *o*, or *ova deficient*, are doomed not because they have inherited faulty genes but because they have inherited faulty cytoplasm. The following are observed: (1) If a female is homozygous for *o*, all her eggs stop cleaving at a certain point and die. (2) All eggs of a heterozygous female, including those with the *o* allele, develop normally. (3) If normal cytoplasm is injected into the eggs of a female homozygous for *o*, development is normal, regardless of the genetic constitution of the egg (Raff and Kaufman, 1983, p. 116). The mother's genotype, insofar as it influences the cellular environment she can pass on to her young, is crucial to their survival, whereas the presence or absence of the mutant allele in the offspring's own genome has no impact on its viability.

   Raff and Kaufman give other examples of maternal effects and go on to discuss the importance of cytoplasmic localization, in which embryological determinants are segregated in differentiation and bring about diverse patterns of gene expression in different cells. Here we have the differentiation of microscopic developmental systems. Similarly, signals from outside the differentiating cells, or even from outside the embryo, play a central role in regulation of development, and many inductive interactions are reciprocal, with cell groups influencing each other's differentiation. Such interactions not only *initiate* sequences of differentiation; they may also be responsible for *maintaining* differentiation in neighboring regions. Development thus proceeds in "cascades" of sequential inductions and "networks" of multiple influences on a given induction; the transience of both competent states and inducing substances is emphasized (pp. 143–147). The authors argue that the redundancy of inductive interactions allows for flexibility in both ontogeny and phylogeny. Ontogeny may proceed despite failure of an inductive event because of multiple causation; if a change in early development is introduced by a mutation, developmental processes may retain their integration and allow the perpetuation of the genetic change (pp. 148–154). Canalization, or developmental homeo-

stasis, is here seen as a result of a system of relationships and their consequences, not of some regulating force emanating from the genome. It
demonstrates that what is *sufficient* in a developmental system may not
be *necessary*. The principle of interaction, in turn, shows us that what is
necessary is not always sufficient.

Raff and Kaufman define evolution by genetic changes (1983, p. 62)
and therefore must, to show that developmental patterns evolve, attribute
those patterns to genetic programs (p. vii). Their "basic tenet" is that the
genes control ontogeny (p. 234). That this is a definitional expedient is
clear immediately; their descriptions are beautifully balanced paragons
of interactionist, constructivist description. Since the ideas of control of
ontogeny by genes and of evolution as evolution of gene pools are absolutely standard (though these authors emphasize regulatory genes rather
than the structural ones used for proteins, and in fact show how changes
in the latter fail to give much insight into evolutionary change in morphology), what is innovative and valuable about their approach is in no
way tied to these definitions. Rather it depends on such things as their
emphasis on the redundancy and flexibility of developmental processes,
what J. Needham called "out-of-gearishness" (cited by Raff and Kaufman
on p. 141), a quality that exists side by side with ontogenetic integration.
This dissociability of certain processes from each other is crucial to Raff
and Kaufman's treatment of morphological evolution by partial uncou-
pling of various aspects of ontogeny. Because they do not deal with traditional nature-nurture issues their definitions raise few problems for their
own treatment, but if their approach is to be adopted outside embryology it would seem that including the developmental system, an extremely
small conceptual step beyond their notions of developmental programs,
life cycles, and developmental pathways, would gain them a great deal
while costing them virtually nothing.

Sometimes, in species that are capable of early dormancy, the conditions for dormant survival and for development are quite different; if the
parent has managed to "pass on" the first but not the second, perhaps
because of climatic changes, reproduction cannot be considered successful. Reproductive success, in fact, will presumably vary with the time and
manner of its assessment. This all suggests that what is inherited is not
definitively fixed at conception but is progressively made manifest in ontogeny; conditions can change and are a partial function of the organism's

own functioning in its niche. This point may seem less bizarre if it is compared with more traditional statements about inherited "tendencies" or about genetic diseases only expressing themselves under certain circumstances. There is uncertainty in such cases over whether one "has" or "gets" the disease. In my terms, one may inherit a particular gene or genes, but it may not become clear for quite a while whether one has also inherited the other conditions for the disease. It follows tautologically that if those conditions appear, then the disease will appear. This is the case whether the genes are "normal" or "abnormal"; if there is no population variability in susceptibility to a disease, the development of the disease still requires both genetic and nongenetic conditions. Burian (1981–1982) and Stich (1975) point out that the distribution of genetic constitutions and environmental conditions determines our judgments about what is a normal reaction to a poison and what is a genetic vulnerability to an innocuous substance.

In one way, a systems view brings biological inheritance into closer correspondence with the model of property inheritance in human societies than has generally been recognized. A family usually passes on its wealth *and* the means for its maintenance and exploitation, including education, social position, and connections, and an appropriate ethic as well. Alternatively, one might say that all those things together constitute the family's wealth. Offspring receiving only part of the complex do not as reliably perpetuate the family fortunes.

If the ecological embeddedness of development and inheritance is construed as the (false!) truism that both nature and nurture are necessary for development, I will not have made my point, which is that all developed form and function, all "nature" (whether we consider it normal, healthy, adaptive, or not), is a function of that embeddedness, and that nature and nurture are not alternative causes but product and process.

Nature is not an a priori mold in which reality is cast. What exists is nature, and living nature exists by virtue of its nurture, both constant and variable, both internal and external. Some aspects of any particular phenotypic nature may be judged pathological. Variation in that pathological phenotype may be correlated with genetic or environmental variation, or both. Some aspects of a phenotype may resemble those of most other species members, and of these, some may show overlap with phenotypes of phylogenetic relatives. Some may be reproductively useful for the individual, and some for its kin or for some other creature or plant, and these are not necessarily mutually exclusive. Others may have been useful at one time but are no longer, because those aspects of the world by which we test usefulness have changed more rapidly than those, genetic and otherwise, that produce the phenotype. Apart from a statement that a phenotype exists, then, the terms "natural" or "normal" have either an ethical meaning, a functional one, a statistical one, or a historical one. The ethical meaning has to do with judgment in a nonscientific domain, the functional one with some criterion of proper workings, the statistical one with relative frequency, and the historical one is essentially an extension of the statistical one to ancestral populations. None of these requires developmental dualism.

2. GOAL DIRECTION AS INTERRELATED PROCESSES

Inorganic forms and processes, as well as organic ones, are causally embedded.[5] What distinguishes a falling rock or the growth of a snowflake, processes Mayr (1982, pp. 48–51) would presumably characterize as "teleomatic" (proceeding by physical law), from those he calls "teleonomic" (proceeding by a program that exists prior to, and is causally efficient in producing, an end contained in itself) is not a program for creating the system but the structure of the system in which the processes occur. Ayala (1972, p. 8) considers teleological explanations to be indispensable to biology but emphatically denies that they require either that end states be regarded as efficient causes or that goals be consciously set and pursued. He defines teleological explanations as those in which "the presence of an object or a process in a system is explained by exhibiting its connection with a specific state or property of the system to whose existence or maintenance the object or process contributes." Such explanations direct attention to the consequences of a process for the system of which it is a part, but those consequences are causally and temporally posterior to the process. The "directively organized" systems to which such explanations apply include both biological systems and certain man-made ones (Ayala, 1972, p. 12; see also von Bertalanffy, 1972).

In another approach to goal-directedness that relies on the structure of the causal system rather than the addition of intention or a program (which is a kind of mechanized intention) to it, Sommerhoff (1974) defines his central concept of "directive correlation" as a goal-seeking activity matched to the environment so as to increase the probability of the goal event above chance. What is crucial here is not that two variables happen to coincide in a useful way (though I would argue that this is much more important in ontogeny than one might think), but that variation in one is tracked by the other, as the possible positions of a grain of food are matched by alternative movements of a feeding bird. The two variables must be "orthogonal," in the sense of allowing for many combinations of initial values (both food and bird may initially be in any number of positions, though their subsequent positions are related), and the goal event is thus their "joint effect" (pp. 74–79). The "coenetic" variable is the adaptively significant one to which the system adjusts over time, which may be brief or many centuries long, as it is in phylogenetic adaptation. Its variants generate a sequence of alternatives in the second, adjustive variable, thus achieving directive correlation (p. 85).

Since directive correlation is defined neither by the reliability of an event nor by its importance for an organism, but by a systematic relationship across a range of variation in both variables (the "domain," Sommerhoff, 1974, p. 105), it is a matter of the structure of causal networks, and cannot simply be assumed from aptness or apparent coherence. What is important about both Ayala's and Sommerhoff's treatments is that they address the question of teleonomic order directly, by examination of the interrelations of events and objects in the world, rather than by metaphor. Sommerhoff particularly warns against making facile assumptions that the brain, for instance, must contain comparators and be controlled by explicit error signals and command signals, just because certain servomechanisms work in this way. He points out that uncritical adoption of machine concepts and reification of input-output relations may encourage fruitless searches for nonexistent brain mechanisms (pp. 10, 103, 120).

According to this view, understanding goal-directedness requires neither mentalistic language nor invocation of machine models, but rather conceptual clarity and investigation of actual relationships among variables and their consequences. Sommerhoff (1974, p. 92) feels that workers in early cybernetics, by failing to formalize goal direction, by becoming immersed in the designing of servomechanisms and then reasoning back from machines to the organisms that were the prototypes for the machines in the first place, increased biologists' reliance on engineering without illuminating biological goal direction.

It would seem that many ontogenetic processes, orderly and useful as they are, would not qualify as directively correlated by Sommerhoff's criteria. This might be because the coenetic (adaptively important) variable does not in fact vary, or because when it does vary within the normal range, the second variable does not track or compensate; in the latter case the variation is biologically irrelevant as long as normal results are produced, and we have simply perceived differences where the organism or its internal processes "perceive" none.[6] When compensation occurs, as is implied by the concepts of equifinality and canalization, the means by which it occurs must be investigated. (Waddington [1960, pp. 81–82] emphasizes that "predesigned end-points" and control by negative feedback should not simply be assumed in these cases, even when developmental trajectories are quite regular.) The appearance of a character in the natural environment of our ancestors or in many contemporary environments is not relevant to the establishment of developmental goal direction in Sommerhoff's system (and goal direction, I suggest, is in some way the intuition underlying some people's eager adoption of canalization as a substitute for maturation), but presence of a coenetic variable and developmental tracking of it *are* relevant, as is a coherent analysis of goals and subgoals that allows identification of directive correlation at various levels. Notice that if compensation occurs, "insensitivity" of result is accomplished by "sensitivity" of process. Insensitive processes may, on the other hand, produce unwonted effects in conditions that lie outside their normal range.

Though Sommerhoff seems to consider only conventional "acquired phenotypic modifications" such as thickening of fur in cold weather or growth of muscles with use as examples of ontogenetic adaptations, it would seem that some of what has traditionally been considered maturation could also fit this category. In any case it would hardly be true that "in the context of ontogenetic adaptations the phylogenetic ones appear as preadaptations" (1974, p. 89). The distinction between ontogenetic and phylogenetic adaptations, that is, is not the old one between two kinds of developmental processes, but rather between a kind of developmental process, on the one hand, and long-term selection of a developmental system, on the other.

Organism-environment adaptation is the focus for most of Sommerhoff's discussion, but he allows for mutual adaptation of variables, giving a tennis game and muscle coordination in walking as examples. In addition he writes of multiple coenetic variables in complex relations (pp. 89–90). Following this lead, we may reasonably extend the analytic scale inward, to some physiological processes, and outward, to include social ones, within and across species. The structural nature of the description allows for units and variables at any level of analysis, including institutional or ecological.

Organized matter interacts in an organized context. Ontogenetic processes may be directively correlated with substances or processes, internal or external, that support and shape change. Internal coherence may be ensured by mutual adaptation. Evolutionary changes in developmental systems involve shifts in these causal meshings, and whether a developing sensory system at some point may be directively correlated with the organism's own behavioral output, with some "external" variable, or with a relationship between the two (Johnston, 1982) is less important for successful development than the fact that the adaptive correlation reliably exists in the organism-niche complex. (Klopfer [1981] suggests that predictability of the pattern of interaction between mammalian parents and offspring may be more crucial for adaptation than the particular style or actions involved.) Similarly, a developmental goal may be reached by linking a developmental process directly to the factors with which adaptive correlation must be accomplished, or indirectly, through a second relation. In the latter case, the required structure or ability can be in place when it is needed; it is a priori in this sense, not in the sense of being static, universal, or outside history.

Johnston (1982), in a treatment of developmental systems as the unit of natural selection, discusses the ecological basis of different developmental strategies found in closely related butterflies studied by Watt and by Hoffmann. In one, wing pigmentation is regulated by day length during the larval stage, while pigmentation in the other butterfly responds to temperature later in development. These differences are explicable by the fact that day length predicts temperature in the environment of the first species but not in the second, and by the thermoregulatory function of pigmentation. After offering several analyses of such variables, Johnston suggests some research strategies implied by an integrated approach to ontogeny. They include investigation of kinds of roles played by developmental influences (following Gottlieb [1976], who discusses the part played by experience in facilitating, inducing, and maintaining development), specificity of effects, varying time scales over which adjustment occurs, and kinds of coenetic variables and their relations. Together "they provide a multidimensional assessment of the relations among different kinds of development, [and] they paint a picture that is too complex for its analysis to be based on any simple dichotomy, such as that between learning and development" (Johnston, 1982, p. 433). One could add to these strategies the specification of levels of analysis and their relations and investigations of the changes in relationships among these influences with developmental state. Underlying such a program of research would be a willingness to view development as a matter of transitions from one structure to the next, and therefore from certain kinds of control to others. Though investigators suddenly bereft of the nature-nurture context (which may function as nest materials did for the infant rats mentioned above) may wish for a replacement, it may be just our assumption that such large-scale simplifying research guides are possible and necessary that needs to change.

This is not to say that selection of variables must be random or that analysis is impossible. It is to suggest that guidance is more likely to come from the system under investigation than from some more abstract assumption about genetic and environmental influences. Fine investigators have always been characterized by good intuitions about what their phenomenon is "paying attention" to; even very close phylogenetic relatives may, as is the case of the butterflies mentioned above, pay attention to quite different things. Scientific talent is partially a knack for reading one's particular system productively.

An appreciation for the way organisms utilize matrices of correlations is seen in Pittendrigh's discussion of adaptive organization as resembling not the work of intelligent design, but rather "a patchwork of makeshifts pieced together, as it were, from what was available when opportunity knocked, and accepted in the hindsight, not the foresight, of natural selection" (1958, p. 400). He describes a certain *Drosophila* fly that must, to avoid dessication, emerge from the puparium as an adult at dawn. The adaptively significant variable is moisture, but it is not moisture but the dark-light transition, which occurs just enough earlier to give the organism sufficient lead time, to which the insect's internal clock is synchronized.

Pittendrigh observes that the language of teleology has been legitimized by cybernetic technology; this is regrettable, he says, because it tends to encourage Aristotelian notions of teleology as efficient cause. It is better, he suggests, to recognize end-directedness by another term. The one he proposes, "teleonomy," perhaps not surprisingly, has often been used to perpetuate just those anthropomorphic causal notions it was intended to avoid.

The limitations of theories that place information in genomes or in the environment are just as marked when we move from *Drosophila* emergence, which depends on the relation between two environmental variables and on the rapidly changing relation between the animal and its surround, to a consideration of the fly's physiological state. In the same way that it is unreasonable to say that information regarding when to emerge is "in" the fly's body or "in" the environment (since light is informative about emergence time only to a particular organism in a particular state), it is unreasonable to say that information for developing a body with just these sequential sensitivities and reactivities is "in" the constituent cells or anywhere else. It is true not only that "information" on emergence schedule is a joint construction of organism and surround but also that information for that state of sensory and behavioral readiness that makes emergence possible is assembled and defined in ontogeny. Again we see that insights from the behavioral level can be applied at other levels as well. We have seen that notions of genetic information are most deeply entrenched in concepts of "physical" maturation and are generally accepted by instinct and learning theorists alike. One may quibble over whether sex roles and aggression are encoded in the genes, but few doubt that sex organs, teeth, and claws are. To refute the idea of genetic (or environmental) coding for mind but not for body is thus not to refute it at all. The view presented here therefore has implications for the relations among the branches of the life and social sciences.

Genes are not just "in" an organism, and organisms are not just "in" an environment. Rather, ontogenetic processes, reliable or not, adaptive or not, behavioral or physiological, are a function of what Johnston and Turvey call the coimplicative relationship of organism and niche (1980, p. 152). P. P. G. Bateson says that evolutionary theory eventually requires an understanding of the constraints on the organism at all stages of its life (1982); relations within the genome, between developing tissues, and among organisms can be seen in similarly interdependent ways. (See Beer, 1975a, on intrinsic selection pressure; Webster and Goodwin, 1982.) To understand an organism is to understand just these relationships, and to understand its phylogenesis is to trace their history. Because the internal and external worlds are highly organized, it is not necessary (nor would it be possible) for evolution to "work on" causally self-contained units, chunks of matter emanating one-way streams of information and influence. Natural selection creates nothing in a conventional sense, however easily it may be thought of as a force or entity. It is a summary of many failures and successes (Caplan, 1981–1982; Nagel, 1979, p. 303). Selection is not an artificer that makes organisms. It does not even make genomes which then make organisms. (The model for natural selection, after all, was not creation, but artificial selection, in which the human functions less like a sculptor than a dating service.) It is the result of many developmental systems' "making" themselves and playing out their few or many ontogenetic interactions, subject to constraints and contingencies at every moment, at every level.

Studies of development and of natural selection are indeed different kinds of studies, but finally they are about the same world. Focus on causes characterizes much developmental research; an emphasis on effects, and effects of effects, illuminates function. But we are looking at the same vital network from two angles, and to grasp the evolutionary history of the network it would seem that we need to integrate the two aspects. It is no longer enough to say that some biologists are concerned with the Great Nexus of developmental causes while others are concerned with the gene as Immortal Replicator, as Dawkins does in an attempt to deal with some criticisms of his approach (1982, p. 99). Perhaps the developmental system, encompassing as it does the inward and outward flows of cause and effect, the relations of organism and surround throughout the life cycle, offers such integration.

Selection occurs at many analytic levels (see papers in Sober, 1984) and at all developmental stages, but variation at any of those levels or times is not statistically random, even though differential survival and reproduction do not directly feed back to regulate genetic variation. The structure of large molecules does not permit "all possible" mutations, crossovers, folded configurations, or interactions. That is, the structure of such molecules partially determines what mutations and interactions are possible. Once a developmental system has attained a certain complexity, physiological functions must be served within certain limits (Stebbins, 1972; Whyte, 1965). One does not have to agree with Whyte's strict opposition of internal selection to "Darwinian external selection" to appreciate his emphasis on developmental and systems considerations in evolution. Selectional pressure, then, takes shape as the developmental system, with all its ecological ramifications, takes shape. Constraints on evolution, that is, are often developmental constraints. To put it yet another way, they are constraints on the possibilities of life cycles. Though constraining and enabling seem antithetical, organization through restriction of possibilities at one level expands higher system capacities (Mercer, 1981, pp. 49–54). Redundancy, the reduction of variety, makes complexity possible, and order is prepared from both inside and outside a developing entity. It is precisely the reduction of variety at certain levels, and the seemingly paradoxical proliferation of complexity at others, that selection accomplishes.

If we give up the gene as unmoved mover, it becomes a variable/constraint like any other. Its mechanisms are no less marvelous for being enmeshed in phenotypic processes; indeed, it is only in the phenotypic arena that its marvels are revealed. The ghost in the cellular machine doesn't make the machine, and it doesn't make the machine run. The cell exists, and it runs "by itself." Part of its running is a matter of gene activation. Another part is its own structure and the context in which that structure persists or changes. The organismic machine is made of cells but its running is not solely a matter of individual cellular programs; its "programs," "rules," and "information" are of a higher level and interact with those on lower levels. All these metaphors for the functioning of the living entity are only that: attempts to convey the nature of vital processes whose order evolves with, and has no existence prior to, or apart from, those processes.

That the ghost is a phenotypic one should not surprise. It is the formalization of our own analyses of phenotypic processes and outcomes. Our mistake is to project it, not back into those processes, but into an originating entity that stands in a creator-creation relation to them. But the genes are not Aristotelian entelechies, nor are they Platonic ideals, essence to mere phenotypic appearance, signal to environmental noise. Nor yet are they selfish little phantoms, scanning the ecological horizon for reproductive gain and operating the levers and relays that make us work, or miniature representations of our phylogenetic ancestors, lodged perhaps in our limbic systems and flooding our nervous systems with animal passions and our ears with irrational whisperings.[7] The ghost must be integrated into the machine, and when the cognitive and causal functions (final, formal, and efficient causes) are wedded with the matter (material cause) that houses them, the necessary synthesis will be accomplished and the need for a ghost will vanish.

If generation is the passing on of an essence, then it appears that what is true to that essence must be explained differently from what departs from it. If matter is passive, incapable of organizing itself, then something else must be called up to assemble it into living beings. And if the essential type manifests itself in tokens despite considerable variation in circumstances, then some force, the teleological synthesis of knowledge and agentive power, must guide the entire enterprise toward the intended goal. These pretheoretical assumptions have many sources, and in the preceding pages I have tried to suggest some of them. The ways we infer intention from a pattern of events (including our own behavior — we sometimes conclude that we must have intended something, or were fated to do it, which is a more metaphysical version of the same intuition, when we see we have repeatedly or persistently brought it about) and think about ourselves as originators of actions and recipients of influences, our philosophical and religious heritages, and the emergence of science from them, all, I think, help form our sense of ourselves as knowers and actors; our mundane epistemology and philosophy of action, in turn, inform our view of other creatures. As subjects, agents, and creators, we find it difficult to imagine how we and our fellow creatures were created, except by agentlike activity. The process of development demanded by these presuppositions must thus provide for knowledge of the type, power to accomplish its construction, and intentional direction of that power. That these are faculties of conscious agents is hardly surprising. Crafting and building have since ancient times been among the predominant metaphors for ontogeny, so it makes sense that formal, efficient, and final causes should originate in the artificer, while material cause should be a function of the raw matter on which he operates (*he,* not *she;* in the Aristotelian system, form, force, and psyche are contributed by the sperm, while matter is supplied by the egg). All this seems with hindsight to be a necessary consequence of our insistence on thinking of development as creation, as *in-formation*.

When the problem of explaining the emergence and persistence of form arises on the biological stage, it is of minor consequence in the end whether the deus ex machina is called upon to create the living machine from the inside or from the outside (or whether there are two of them, each contributing certain components); the yoked ideas of preexisting form and teleological control travel together. They require each other, imply each other, even conjure each other out of the wings if separated, exchange places in successive performances. If form is latent inside it is innate, if it is impressed from without it is acquired; maturation and motivation supply the guiding forces. The re-creation of the creator inside or outside confers the same basic but finally unsatisfactory benefits as the ghostly causal will does in Ryle's account. It feels right, but it doesn't explain anything. It simply repositions the mystery.

I have described the ways in which I believe that these maneuvers, while often motivated by the need to account for some empirical result or other (developmental regulation, behavioral variability, species-typical learning, etc.), are ultimately false to the very facts we report. It is we who must switch codes, so as to do justice to the world revealed in our analyses, to interpret them productively, and to avoid dead ends in our research and theory. Organisms are not simply more or less faithful representatives of an internal species archetype or of an internal individual plan, though some of the evolutionary theorists who have argued most forcefully for population concepts and against essentialism ensure, in the language of programmed development, the perpetuation of a kind of essentialism. Matter is not inert, though the very molecular biologists who have most effectively shown us its exuberant, irrepressible (usually) reactivity have often insisted that information must organize masses of matter. The cell is not a bag of chemicals, though the investigators who have the deepest appreciation of the complexity of its activity and its structure sometimes speak as though it were. Nor, surely, do these people believe in inner essences, in the inertness and inchoateness of matter. Their conceptual systems require, however, that whenever they shift from description of processes to general explanations of those processes, the agent-patient, informer-informed relation be reinstated. As long as coming into being is understood as the instantiation of preexisting form, the fundamental aspects, and therefore the fundamental contradictions, of our ways of construing life will reconstruct themselves into new versions of the homunculoid gene as surely as a sponge remakes itself after being put through a sieve.

Viktor Hamburger (1980), in a paper on the contribution of embryology to the evolutionary synthesis, points out that once they began to study proximate causes of development, embryologists paid little attention to evolutionary issues. Geneticists' emphasis on the role of the nucleus in development and embryologists' concentration on the cytoplasm hindered communication. So did the latter's concern with individual and intra-organismic processes, in contrast with the focus on populations seen in the modern synthesis. Hamburger points out that what was required to transcend the nucleus-cytoplasm opposition was some sort of "nucleo-cytoplasmic interactionism," aspects of which he detects in the thinking of a few students of development, including Schmalhausen and Waddington. The ideas of these last two showed how "epigenetic mechanisms" could "take a great burden off the genome in its role as controlling agent of developmental processes" (pp. 103, 198).

It would seem that until these constructivist interactionist ideas are clarified, assimilated, and applied across not only the nucleus-cytoplasm boundary but across progressively higher ones as well, the burden of agency will not be entirely lifted from the gene, and the integration of embryology and other developmental studies with evolutionary theory will remain incomplete.

In this chapter I will summarize my argument by recapitulating several themes and by showing how each tells us something about development and gives us guidance about how to think more effectively about it. I will also attempt to respond to what I imagine to be the final promptings of the still-skeptical reader, who, having voiced objections in earlier chapters, has nevertheless been faithful enough to accompany me thus far, who finds merit in my discussions, but who is still uneasy about really giving up the nature-nurture complex.

The developing organism as object of thought will occasionally be compared to the process of thought itself. Our conceptual structure and metaphors not only describe our discoveries, they guide and define them as well, and both object and knowledge of it emerge interactively. There may also be a nonarbitrary relation between the fact that we find it difficult to believe ontogeny is possible without a guiding mind or mind-surrogate to lend impetus, direction, and form to the process and the fact that we find it difficult to conduct our lives without recourse to a priori truths, particularly in the face of social and cultural variety and great uncertainty about the future. It is this relation that leads us not only to place God in the cell to make us, but to search ontogeny and phylogeny for clues to the enduring conundrums of the place of humanity in the world, of fate, reality, and limits to human will.

A. Levels and Essences

The "informational" significance of any developmental influence, as we have seen, depends on the state of the entire developmental system, including genes, the rest of the phenotype, and relevant aspects of surround, and on the level and the type of analysis. Developmental state is a kind of temporal slice through the life cycle. It carries the evidence of past gene transcriptions, mechanical influences inside and outside the organism, results of past activities, nutrition or the lack of it, and so on, and it has certain prospects for change. If we are guided by the notion of information as the difference that makes a difference, then what developmental interactant makes a difference depends on what is developing, and how. Understanding ontogeny thus becomes partly a matter of charting the shifts from one source of change (including intraorganismic processes) to another, as one interaction alters the developmental system in a way that provides transition to the next. Equally important are the means whereby stability is achieved. In addition, the organism can be investigated at the behavioral, physiological, or other level, and comparisons made between a system and itself at an earlier time, among members of the same species, between a species and its phylogenetic ancestors or its contemporary relatives; again, what makes a difference depends on what question is being asked.

Some traditional nature-nurture questions can thus be restated in terms of developmental systems. The benefit of this restatement is that it makes clear what is really being asked, and therefore what would constitute an answer. The mischief of the nature-nurture complex was that it conflated both questions and answers, so that an apparently spontaneous change in behavioral pattern, for example, was attributed to the genes, and thus concluded to be unchangeable in the individual, to be universal in the species, to show cross-species identity. Conversely, a cross-species resemblance was taken to mean that no learning had taken place, that no learning would have an impact, that the trait was more real than others that did not show such resemblance. To gain information we need to specify a context and a set of possibilities. It is in this sense that organisms generate information (Klopfer, 1969, 1973, p. 27), and it is in much the same sense that scientists do. Events do not carry already existing information about their effects from one place to the next, the way we used to think copies of objects had to travel to our minds for us to perceive them. They are given meaning by what they distinguish. Thus we find that a gene has different effects in different tissues and at different times, a stimulus calls out different responses, including no response, at different times or in different creatures, and an observation that is meaningless or anomalous at one stage of an investigation or to one person becomes definitive under other circumstances. A difference that makes a difference at one level of analysis, furthermore, may or may not make a difference at another. This is, in fact, the key to understanding apparent spontaneity. A view of the biological world that reduces cause to discrete genetic and environmental forces reduces living beings to infinitely thin membranes resonating to signals from within or without but lacking the substance to generate signals of their own. Seeing the organism as layered vital reality, on the other hand, allows us to see that lack of change at one level may conceal change in the dynamics of a lower one. In the same way a cell may be embryologically determined before it is visibly differentiated, a mind may be in a state of readiness before it processes in a visibly new way. This is not because the cell was always determined, and it is not because the mind always had knowledge of what it would encounter. It is because the way a cell or a mind engages the world is a function of its structure, a structure that emerges from activity at multiple levels. The changes in a female human at puberty seem spontaneous until one looks at the cycling of hormones, for example, which change over the prepubertal years; and they seem fated until one looks at the ways in which the hormonal system is linked to the outside world and the ways developmental schedules can be influenced.

Though this description may seem to evoke the very notion of hidden reality against which I have argued so heatedly, it is different in an important way. Reality is differentiated at its various levels, and the phenomena at one level are not simply reduced or incomplete versions of the phenomena at another. Sober (1983, with nods to Attneave, Dennett, and Fodor) defends a mentalistic psychology, and in answering the behaviorist criticism that such a psychology simply posits useless homunculi, points out the importance of positing *stupid* homunculi. It is homunculi that simply contain the phenomenon to be explained that are useless. If, on the other hand, they allow us to explore levels of processing that eventuate in the phenomenon to be explained, they can be useful indeed. This is fertile ground for interdisciplinary studies, but even within a discipline such a perspective is imperative, for it allows an investigator both to appreciate what is unique about his or her subject and to connect it to the processes that generated it without reducing it out of existence. What is "given" need not be timeless. Biological givens, the structures that allow function, are indeed given by time, but time's arrow is the vector of change and the emergence of structure. Our world is one of imperfect correlations, whether one is mapping diachronically or synchronically. Its intricacy would hardly be possible without such loose-jointedness.

The developmental meaning of any influence thus depends not on its carrying "information" in some absolute sense, but on what it is related to in the larger developmental context, just as the scientific value of a datum depends on its empirical and theoretical relations; this is why inference is as important as observation. Gottlieb (1976), for example, describes experiential effects as facilitating, inducing, and maintaining development. P. P. G. Bateson (1983) adds a predisposing or enabling role to the list and makes the important point that these descriptions may be applied to internal influences, including gene products, as well as external ones. (Not that experiential influences are necessarily external.) In an example of this kind of unified analysis, Rosenblatt and Siegel (1981) describe the transition from hormonal to sensory regulation of maternal behavior in certain mammals. (Note that switching from the physiological to the psychological levels does not mean that the latter processes are not also physiological.) Lest one forget that hormones are responsive as well as responsible, the same authors describe the importance of the fetal placenta in regulating estrogen secretion by the mother's ovaries.

Shifting among levels of analysis is not to be conceived, then, as movement along some axis defining degrees of essential reality or causal primacy, but rather as choice of dimensions along which a complex reality is constituted.[1]

Our understanding of development, like "developmental information," is dependent on the state of the inquiry and on the level on which it is conducted. As we move down the levels, we do not approach the True or the Real or even the Rockbottom Causal. We do change scale, vocabulary, concepts, and method. Often we find that both ambiguous and degenerate (one-many and many-one, not to mention many-many) mappings exist between levels. It may even be that the existence of both ambiguous and degenerate relations marks a shift in levels. A process at one level, that is, may mean more than one thing at another, and several processes at a "lower" level may specify the same one at a "higher" one. In linguistics, the demonstration that sentence synonymy and ambiguity could not be explained by a single-level grammar was one of the things that signaled the need for a new approach to language.[2] In ontogeny we find that a single genotype may be developmentally mapped onto many phenotypes, not necessarily in continuous variation (consider certain insect castes). In addition to this ontogenetic ambiguity we have the phenocopy, a sort of ontogenetic synonymy whereby different genotypes, developing in different circumstances, converge on the same form. A given arm movement may have many possible meanings, while a variety of movements may constitute the same act.

For coherent integration to be accomplished, an investigator must do by will and wit what the developing organism does by emerging nature: sort out levels and functions and keep sources, interactive effects, and processes straight.

Our concepts of information and knowledge have a complex history of their own. If science is not an interaction between God's mind and our own, but rather one between our minds and the world, then our analyses of the world are not a matter of deciphering some preordained fate or a set of prescriptions, proscriptions, and ultimate constraints inscribed in DNA, but rather a way of encountering the world's multilayered complexity. Moving "down" the layers is not moving from effect to cause, from the provisional to the immutable, from the trivial to the profound, from the merely mental to the fundamental.

· Clifford Geertz (1973, pp. 38–39) describes the "stratigraphic" concept of these layers: man as "a hierarchically stratified animal, a sort of evolutionary deposit." To find the essence of man in this view is to strip away the variable distortions of the upper levels to reveal the universal invariant nature postulated by Enlightenment thinkers: "the naked reasoner that appeared when he took his cultural costumes off." (Though the naked reasoner has in more recent times sometimes tended to become the impulsive naked emoter, while reason is seen as thin overlay, the stratigraphic vision tends to persist.) Geertz speaks of a conceptual "halfway house between the eighteenth and twentieth century," the idea that some of culture is universal and a function of inner Newtonian forces.

A more recent partway house, one that Geertz does not mention in his paper, bisects the distance between that earlier midpoint and the end by dividing universals themselves into the biologically and the culturally determined. Is there an infinite series of Zenoic solutions between this three-quarter-way house and the end of the course?

Having argued that human essence is revealed as much in its variations as in its constancies, Geertz notes that anthropologists turn to universals from a fear of losing their bearings in a culturally relativistic world. I think there are several ways in which this is true. Cultural relativism, in part an expression of liberal distaste for ethnocentrism, was rudely challenged by the events of World War II, which seemed to leave anthropologists without an intellectual and ethical platform from which to justify their opposition to Nazism. Some had tried to derive a kind of naturalistic ethic of tolerance from cultural diversity itself (Benedict, 1934, for example), but this did not provide the means for confronting a world that was rapidly changing from a collection of self-contained cultures to a shifting landscape on which Volk decimated Volk on a scale that defied comprehension. Loss of intellectual bearings is as uncomfortable as loss of moral ones, and indeed, the two are hardly independent. Many anthropologists, no less than other students of life, are struggling to construct a way of moving beyond stratigraphic analysis—which, not coincidentally, stratifies scientific turf as well as ontological levels, a fact that has not facilitated synthesis. An armed raid across disciplinary boundaries is as difficult to view with equanimity as one across geographical ones. Ultimately I am quite sympathetic with E. O. Wilson's desire for a biological synthesis. I just think he has gone about it wrong, and his rather undiplomatic rhetoric, in which one discipline cannibalizes another (1975, chapter 27), is not just a tactical blunder. It is intimately tied to the way the disciplines and their relations are misconceived.

Behavior is not booty to be borne off by those who execute or successfully subvert reductive analysis. A demonstration of variability or mutability does not remove a phenomenon from biology, any more than a demonstration of constancy removes it from the social sciences. Such skirmishes, of course, are hardly restricted to these particular frontiers. A Great Man interpretation claims a character for psychology, while a historical determinist one snatches him back into the surge of supraindividual forces.

The growth of our own understanding seems to me to require a reworking of our notions of disciplines, causality, and explanation, not because I have any particular reverence for turf claims but just because I do not. People *will* find a way to study what intrigues them. Brief inspection of the lists of participants of recent conferences and workshops on behavioral development will reveal enough turf hopping to give a departmental conservative the vapors. But we are hindered in our attempts at productive cooperation by just the assumptions about levels, phenomena, and causes that are usually used to define disciplines. *Is* it the case that if something is too complex to be acquired by learning as learning is presently conceived, then it is properly studied by embryologists and physiologists, not psychologists? Or should the provinces of those fields, as well as our assumptions about maturation and learning, be adjusted? What kinds of information count as information on what kinds of processes?[3]

What is the relation among the vocabularies, concepts, and methods of the various sciences? What kinds of constructs are appropriate to what kinds of data? Is reduction the only interesting bridge between disciplines (and is preemption a bridge)? Can we develop a new set of relations that will better serve integrated inquiry? Mightn't such a set allow for mappings among levels, sometimes very accurate but more often imperfect, rather than the "nothing but-ism" and "real cause-ism" that characterize much "interdisciplinary" work at present? Rose (1981) considers causal language inappropriate for relations between levels; correspondences may be found among levels, but this is logically distinct from causal sequences, which can be identified within levels only. Levels require different languages and methods and allow us to approach phenomena in a variety of ways.[4]

I can hardly decide for the rest of the scientific world how the biological is to be defined, and maybe I don't have a totally satisfactory answer for myself. We should recognize, though, that this is an important question of definition and strategy. In the course of a chat about a theory involving certain behavioral correlates of autonomic nervous system (ANS) function, a colleague mentioned problems with the interpretation of the research on heritability of variation in ANS reactivity. He concluded, "I'm not convinced it's biological." I was struck, not because there was anything nonstandard about his comment, but because nothing could be more biological than the ANS, wet and slippery as it is. Yet our vocabulary allows us to say that its degree of reactivity may or may not be biological because, among other things, genotype and environment were imperfectly separated in some research. That this is not a simple case of lexical ambiguity is, I trust, clear from preceding chapters.

Immelmann et al. (1981, pp. 3–4) comment on the dual use of "biological" as "physiological" and as "innate or instinctive," and on the difficulty of keeping these usages separate. (This is, of course, both more and less than dual use, because of the multiple meanings of "innate" and because the innate is often defined as physiological.) They note the problems encountered in their attempt to forge a common vocabulary at the Bielefeld Project and propose a different meaning for the biological: "shaped by natural selection and . . . therefore adaptive in the environment in which it evolved." This alternative meaning is presented as "interactionist," not genetic determinist, and as explicitly teleonomic. I am not convinced that this formulation resolves the problem, though I appreciate the attempt (realizing something is amiss is half the battle). As we have seen, such definitions tend to collapse fairly easily into a form of preformationist determinism by way of preprogrammed epigenetic rules, genetically guided canalization, phylogenetic information in genomes, and so forth. Certainly some of the things that need to be elucidated are what it means for behavior to be shaped by natural selection, how one knows when behavior has not been so shaped, precisely how one recognizes adaptiveness, and what the relations are among past and present selection, adaptation, and normality, particularly when dealing with a species, like our own, that rarely inhabits the environments in which it evolved.

Though I absolutely agree that the biological needs to be redefined, my own sense is that we will be better off in the long run if we identify it either with a level of analysis roughly equivalent to "physiological-morphological" or all-inclusively, as "the science of life." In the latter case it will in some sense embrace the behavioral sciences, since we all study the workings and doings of living beings. Because new kinds of relations and phenomena emerge at each level, however, fields are not necessarily theoretically reducible to each other, and academic imperialism is not justified.

Defining biology by adaptation, on the other hand, leaves nonadapative physiology and morphology without a discipline (unless all physiology and morphology is defined as adaptive, which creates problems of its own) and slices the other levels of analysis into biological and nonbiological sectors. This in turn is both supported by and supports a dual notion of ontogeny, which is precisely why predetermined epigenetic rules slip so smoothly into place and why thoroughgoing constructivist interactionism is so unstable—not because it is wrong but because it is incompletely integrated and thus drifts rapidly toward the traditional poles.

The resistance to deep conceptual change described in these pages has some of its roots, I suspect, in a fear of relativistic vertigo, both moral and intellectual. Phylogenetic ordering is seen to fix moral forms, functions, and limits to save us from ethical chaos, while it simultaneously brings order to a disorienting landscape in which instincts are sometimes learned or may go awry because the wrong thing was learned, where docile lab animals are moved to insubordination by "instinctive drift," [5] or show species-typical biases in conditioning experiments, where "lower" organisms apparently perform the most astounding probabilistic calculations and "higher" ones behave in the most beastly manner imaginable.

We are engaged in the most precarious sort of bootstrap operation, attempting to correct the very scaffold on which we stand. It cannot be otherwise. Which tools to jettison? Which intuitions to ignore? How to judge new ones? How much can we discard before we find ourselves without the means to stand, let alone build? (I am not blind, incidentally, to the irony in using a construction metaphor to describe the development of ideas. There is a difference, though, between having one's ideas constructed by genes or by stimuli and constructing them oneself, within one's particular life contexts.) There is not, however, despite my own emphasis on conservatism in interpretive systems, only one way to construe the data of our senses (or I would not be proposing an alternative), just as there is not only one developmental system that can surround a genome (or there would be no norm of reaction).

B. Alternative Pathways

Phenotypic variation is not just different ways of *appearing,* it is different ways of *being.* Whether one speaks of variation or its absence, it is the developmental system (or, to avoid the considerable temptation to reify a system, the processes that collectively and successively constitute the developmental system) that integrates inner and outer, past and present, regulating the utilization of genes, foodstuffs, and stimuli that may be continuously present but intermittently effective, helping to generate its network of support and stimulation as it does so.[6] At all points it is intensely selective, becoming unresponsive to certain influences but profoundly affected by others. The point of interactionism is not that everything interacts with everything else or that the organism or genome interacts with everything. Nor is it that everything is subject to alteration. It is rather that influences and constraints on responsiveness are a function of both the presenting stimuli and the results of past selections, responses, and integrations, and that organisms organize their surroundings even as they are organized by them. This being the case, developmental pathways are not set in any substantive way either by genome or by environment, regardless of the normality or relative probability of the pathway itself.

I suggest that Waddington was, in his genetic assimilation work, at least as concerned with alternative developmental pathways as he was with buffering or canalization of those pathways. (He was certainly not concerned with the transformation of acquired characters into inherited ones, but with challenging the conceptualizations of development that require that distinction in the first place.) Recall that he was first drawn to the idea of canalization by alternative developmental pathways of cells, that is, by tissue differentiation. It is true that he was struck by the absence of grading between tissue types—hence the notion of buffering. But the basic fact is that distinct tissues arise from the same zygote. Similarly, in the genetic assimilation work, the primary observation was that a creature developed in ways that depended on its circumstances, and that, far from being irrelevant to evolution, such ontogenetic variation could on occasion be crucial to it. One could sometimes select for an ability to develop in a particular way under unusual conditions until the developmental pathway had changed and could be negotiated even under the original conditions, conditions that had not previously given rise to the character in question. The fact that the genetically assimilated character now appeared in the absence of the unusual environment that was originally required for it is consistent with the notion of genetic characters as spontaneous and independent of conditions, but to interpret the findings in this way is to miss the point. The point is that there are several ways of developing that character, each way environmentally and genetically contingent: (1) the original genotype in the altered circumstances, (2) the selected genotype in the altered circumstances, and (3) the selected genotype in the original circumstances. One can, furthermore, sometimes move by selection from one way to another. This multiplicity of developmental possibilities gives the species grace time to adapt genetically to new conditions while at the same time giving selection a new process on which to work.

Similar points have been made with respect to the role of behavioral adjustment in evolution. It may allow the animal to survive in an altered environment, at the same time bringing new internal and external selective pressures to bear. One might consider such behavioral adjustment to be a special case of development along alternative pathways, a process that goes on not only at the behavioral level but within the organism as well. A tissue that usually develops in a certain way may, because of altered developmental circumstances, develop in a different way. This may have beneficial consequences (or be neutral) and may also expose the organism to new pressures, by putting more mechanical stress on a bone or by competing with another organ for nutrients, for example, creating new ontogenetic problems to be solved by further changes in the developmental system. Evolution works not only on alternative finished phenotypes (when is a phenotype finished?) but on alternative ontogenetic pathways as well, at organismic and intraorganismic levels. Phenotypic variation, furthermore, on which selection depends, can involve both genetic and environmental variation. It may be because behavior is associated with malleability and variability, while morphology and physiology are seen as resulting from rigid "maturational" processes, that the role of behavioral variation in evolution seems more readily accepted than alternative routes at the chemical or organ level. They seem nevertheless to be conceptually similar. While Waddington experimentally altered his subjects' environment to bring about the initial increase in phenotypic variation from which he made his selection, in the natural world organisms may progressively alter their own environments (perhaps irreversibly) or move into different ones; it is conceivable that by doing so they could set off an analogous chain of events and consequences, contributing to the evolution of their own lineages.

Like cells, tissues, and developing organisms, we as thinkers often have both fewer and more possibilities than one might think. Some inferential routes we have habitually taken can be shown to be dead ends, to converge or to loop back on themselves. But sometimes data can be interpreted in more than one way, and while this can be a nuisance, it can be our salvation, too. Also like developing organisms, we often expose ourselves to new implications, pressures, and challenges by varying our ways of intellectually engaging the world.

Shirley Roe has described the changes in the thinking of Albrecht von Haller (1708–1777) from spermaticist (animalculist) preformation to epigenesis to ovist preformation. These shifts often involved reinterpretation of the same observations and the redeployment of similar arguments for different purposes. Roe believes that the theological and philosophical implications of various views of generation (reproduction and development) exerted a powerful influence on Haller's thinking, as well as on that of his contemporaries, including the young Caspar Wolff, the rationalist epigeneticist ("epigenesist" in Roe's spelling) whose debate with Haller she chronicles.

They were arguing, Roe says, about "how one *ought* to explain embryological development," and about the criteria for scientific explanation — in short, about the philosophy of science (1981, p. 88). The proper places of God, chance, matter, and natural forces were the central concerns, and in their modern guises, they are ours as well. In 1757, faced with Georges Buffon's ideas of self-directing matter and immaterial vital forces, Haller "came face to face with the problem of how development is organized and directed" (Roe, 1981, p. 43), a problem he solved by embracing preformationism. It is a puzzle that faces us still, and that we persistently attempt to resolve in ways that bear a certain resemblance to the ones that historians of this period describe. To Haller the possibility that matter could act on its own threatened the role of God as creator and ruler of the universe. He maintained that it was dangerous to admit the formation of even a single finger by other than divine power, for this would eventually allow self-creation by man himself (p. 98).

Wolff, by contrast, explained development in terms of an essential force by which various kinds of matter exerted mutual attraction and repulsion. By allowing matter to act, he avoided the necessity of positing preformed structure and divine forces. He believed that simple forces, in contrast with vitalistic building forces, have by nature but a single action, but that by operating in different situations, they can have different ultimate effects. By developing in a variety of circumstances, he wrote fairly late in his career, a species could have a variety of forms, *none of which was more essential to the species than another.* The species (type of substance), however, did not change, but was perpetuated regardless of the particular form developed. Constancy of structure was attributed to constancy in environmental conditions as much as to the substance. Structures themselves were not signs of the species, as the taxonomists believed, because they were all potentially alterable. Though Wolff believed in the immutability of species, he seemed to be more impressed with variability as such than many other biologists of that period or later. (Despite the fundamental importance of variation to evolutionary theory, variants themselves have tended not to be the focus of interest; see Bateson and Klopfer, 1981.) His was an essentialist view, but it differed from those of most contemporaries in that he refused to identify any of the visible forms with the invisible material that defined the species (Roe, 1981, pp. 130–147).

Is it possible that Wolff's emphasis on multiple developmental possibilities and his unwillingness to consider some characters as more expressive of inner reality than others were simply not congenial to the emerging nineteenth-century interest in the relationships among species, in phylogenesis as evidence of the universe's basic tendency toward growth and progress, and in uniform developmental courses as evidence for both? To this nonhistorian, Wolff's ideas on developmental variation, as well as his attempts to avoid both mechanistic reduction and vitalism, were impressive and fascinating, standing in striking contrast to a background of thinkers, epigeneticists and preformationists alike, who were preoccupied with single developmental routes and, aside from the occasional anomaly, inevitable outcomes. He is seen by Roe as something of an oddity, a thinker whose ideas did not have much influence on the coming Kantian biology of "generic preformation" and "teleological epigenesis" (1981, p. 152).

Like our eighteenth- and nineteenth-century predecessors, I believe, we have more conceptual possibilities than we have exploited. Far from wiping out our store of analytic tools, the realignments I suggest here should free latent variation as artificially linked concepts are decoupled. Reliable, seemingly spontaneous and stable developmental sequences can be studied as closely coordinated systems of interactants, without inner essences, whether Wolffian, preformationist, or modern. Their qualities result not from insulation of essence from the world, but from the re-creation of species form, common and variable, from systems that tend to perpetuate themselves by virtue of innumerable interactions at all developmental stages and at all levels.

The importance of phenotypic variation in the absence of genetic variation is not that it shows some character to be only partly encoded. Such interpretations only reinforce the belief that *lack* of variation is due to agentive genes. Even the controversy over biological versus cultural universals depends on the homunculoid gene; otherwise there would be nothing to contrast cultural universals *with*.

One salutary consequence of dropping our inference to, and explanation by, information in genes or in environments is that developmental continuity, stability, and uniformity, as well as discontinuity, lability, and variability, can be given the theoretical and methodological scrutiny they deserve rather than being imputed willy-nilly on the basis of nature-nurture assumptions.[7] A chick may not have to "learn" to gape when it sees its parent, but neither does it gape all its life. The most productive reaction to too many assumptions of continuity, however, is hardly a declaration of a new overriding principle of plasticity; if it is treated as a principle, it promises nothing but another ride on the pendulum. Rather, what is needed is a deeper inquiry into the ways developmental processes are regulated.

Disciplines should be seen as complementary enterprises, not in the sense of carving up a phenomenon into, say, phylogenetic and adaptively modified components, but in the sense that they offer alternative paths to understanding. This does not mean that there is no relationship among levels. The point of multilevel analysis is not, though, to engulf or defend territories; it is to coordinate inquiry and to seek mutual intelligibility. Grmek (1972) observes that chemical and physical interpretations of life were opposed to each other for a long time before they were accepted as complementary.

I do not mean to imply that the question of identifying legitimate levels is a simple one. (See discussions in Greenberg and Tobach, 1984.) As is the case with identifying adaptations, we tend to rely heavily on intuition, and intuitions are notoriously variable. Perhaps academic precedent is no better as a guide. The problem of clarifying the concept of level, however, has been exacerbated by many of the conceptual and ideological issues I have reviewed; increased clarity on these matters may not solve the problem, but it should increase our chances of making real headway.

Thinking in alternative ways about the many ways development can proceed, then, seems to require thinking about thinking in different ways as well, and this includes rethinking inferential habits and even disciplinary boundaries, not because the boundaries are so important in themselves, but because they indicate and encourage ways of allocating causes and explanations that may be more hindrance than help in achieving interdisciplinary cooperation.

C. Ecological Embeddedness and Inheritance

I. EVOLUTION AND DEVELOPMENTAL DUALISM

The traditional view of the relation of development to evolution is that a species adapts to a niche, which itself may be changing, by genetic alterations that allow organisms to mature into adapted phenotypes. Features formed by "nongenetic" processes may or may not be useful to the organism or its offspring, but they are not passed on and thus do not contribute to evolution. The role of behavior in influencing selectional pressures is a refinement but doesn't necessarily challenge the underlying assumption that inherited and acquired characters develop in different ways, and that, indeed, one of the defining features of inherited characters is that they do not develop in response to particular aspects of the surround, but are formed from within while being supported and nourished from without.

The adventures of the emerging phenotype, that is, are almost by definition irrelevant to the evolutionary process, their only role being to land the organism in reproductive condition at the apex of its biological career.

Recently, renewed attention has been focused on the part played by ontogeny in phylogeny, a topic that has interested a variety of scientists over the years (see de Beer, 1958; Gould, 1977; Raff and Kaufman, 1983, for some historical discussion; see also Waddington, 1975; Whyte, 1965). Though I obviously applaud this resurgent interest, I maintain that we must fully relinquish developmental dualism in order to integrate these two long-alienated fields, and that fuzzing the line between the innate and the acquired or turning the dichotomy into a continuum of degrees of programming or indirectness of genetic influence (Anastasi's 1958 solution) will not do. Though anyone, if pressed, admits that any "genetic" character requires the proper environment to appear, this is quite beside the point, since, as we have seen, the idea of dual processes relies not on genes or environment as sufficient causes but rather as alternative sources of form.

The usual demonstration that "acquired" characters are not "inherited" involves following a fairly pure strain through one or more aberrant life cycles, which produce aberrant phenotypes, and subsequently returning it to the original conditions. Naturally, since the original developmental system is usually reinstated, the original phenotype usually appears and the unusual phenotype is pronounced "acquired." If the aberrant life cycle becomes the typical one, however, the conditions for the new character will be present for each new generation. If the Grim Tail Chopper of Chapter 4 appears only once, short tails do not become a feature of mouse life, though the result of the trauma is as much a function of the mouse's genome as anything else it does or is (mice can survive such amputation, they do not regenerate tails, etc.). If the Chopper is a stable feature of the habitat, so are short tails, assuming the developmental system as a whole is a well-functioning one. The Chopper can be reliably present in the developmental system because the mice seek him, because he seeks them, or because he simply lives where mice live. If Choppers are irregularly distributed over mouse habitats there will be morphological variation, and this variation may be correlated with genotype if, for example, mice tend to nest in areas similar to or close to the ones in which they were reared. We do not have to invoke a genetic propensity to get one's tail shortened or to respond to a cleaver to show why mice have short tails, though in this sort of interaction one can readily imagine all sorts of selectional pressures on various aspects of the system, depending on the adaptive significance of being chopped and of chopping for the two organisms involved. Dawkins tells us that queen ants in many species use their wings only for the nuptial flight, after which they bite or break them off before going underground (1982, p. 42). No genes distinguish the winged ants from the wingless ones; an ant grows wings only if raised as a queen. In this case the counterpart of the Grim Chopper is the Grim Chomper, and Chomper and Chomped are one.

Given geneticists' emphasis on the gene as source of order and embryologists' on the cytoplasm, emphases that Hamburger noted in describing the relative exclusion of embryology from neo-Darwinism, and given the importance of the doctrine of one-way information flow in distinguishing legitimate neo-Darwinian from illegitimate "Lamarckian" conceptions of evolution, it is perhaps not surprising that synthesis has been so difficult. Raff and Kaufman give an interesting account of the various "divorces" among the fields of molecular genetics, Mendelian genetics, and developmental biology, as well as the fate of various attempts at bridging the nucleo-cytoplasmic gap (1983, chapter 1). The embryologists' reasons for rejecting geneticists' views, which included the preformationist implications of particulate inheritance, the need for cytoplasmic direction to account for differential gene activation in differentiation, and the neglect of trait elaboration in the life cycle in favor of trait transmission across generations, are still important. They are unlikely to be met fully until we give up the idea that traits are transmitted.

2. EXTENDED PHENOTYPES AND EXTENDED DEVELOPMENTAL SYSTEMS

When Dawkins (1982) speaks of gene effects spreading far beyond the boundaries of the organisms in which the genes are housed, influencing other organisms and the inanimate world in the unbounded extended phenotype, he inadvertently supplies some persuasive arguments for the notion of the evolution of developmental systems. He emphasizes the evolution of genes for their ability to "cooperate" with other genes, including genes in other bodies. So if the "living world can be seen as a network of interlocking fields of replicator power" (p. 247), it must also be a network of interlocking fields of dependence. If a gene "controls" and "manipulates" the world, it does so only because it "needs" all these extended phenotypic manifestations for successful replication. (And we have already seen that control must itself be controlled to be effective.) All these artifacts and organisms must play their roles in completing the jigsaw puzzle of the developmental system, and Dawkins makes an important point when he emphasizes the dispersion, the discontinuity, of these developmental interactants.

To adopt Dawkins's gene's-eye view for a moment, we can see that it would make sense for a gene to take advantage of any developmental opportunity, without caring whether the influence originated inside its organism's skin or outside it. Viewing this widely ramified network of interactions in terms of extended phenotypes rather than of developmental systems, however, has several disadvantages. First, if a gene's phenotype may be part of another organism's body, then any organism's genotype would seem to be distributed as well. Just what genes were part of that genotype, furthermore, would change with time, since different genes would "manipulate" this particular body at different times. Second, even if one retains a more mundane view of genotype roughly as that complement of genes enclosed within the skin, the organism in Dawkins's account is not only something of an epiphenomenon to genetic wheelings and dealings (as it already seems in many sociobiological accounts), but a mosaic epiphenomenon to boot, created and run by its own genes and by the genes of multiple others. The concept of the developmental system, on the other hand, incorporates the insight that a given phenotype is a product of quite a bit besides its own genes without doing away with the individual organism itself. It is ironic to me that biologists who begin by being enthralled by the forms and workings of plants and animals sometimes end up analyzing them out of existence.

3. UNPACKING THE DEVELOPMENTAL SYSTEM

Many are ready to admit problems with the concept of the innate, but there still seems to be a set of characteristics that require special explanation—features that appear very early; that seem stable and fundamental to individuals and/or to species; that are closely tied to developmental state; that are immediately useful, specific, and effective without being learned; that seem to link us to our phylogenetic relatives. Some a priori basis seems needed to explain such phenomena. The continuing conviction that they have unique developmental status makes it hard to make any more than superficial adjustment to the objections of those who are troubled by inconsistencies in the conceptual foundations of thought about biological nature. I share the conviction that these aspects of organisms are important, and do not think it helpful to wave them aside by saying that they are significant in lower forms but not in higher ones or by relegating them to embryology and physiology. The first strategy simply invites arguments about relative numbers of instincts while doing nothing to clarify the concept, while the second confuses levels of analysis with kinds of development. The qualities listed above don't necessarily go together, though their association in a "biological" bundle tends to be assumed, so that one is freely inferred from another (Beer, 1983a; Oyama, 1982). Nor, it is increasingly being realized, do they necessarily form a useful contrast to learning, which is also strictly constrained and directed by existing structure and which may be very early, fundamental to species and individual characteristics, and so on.

One thing that clarifies these issues is distinguishing questions that have to do with developmental state (early appearance, presence at birth or before exposure to "appropriate" stimuli, selectivity, or specificity) from those that have to do with uniformity within individuals (cross-context stability, stereotypy) or across individuals (universality), and keeping these separate from the question of levels of analysis (biochemical, psychological). Phylogenetic relatedness is a historical issue that often masquerades as a developmental one.

The first type of question must be approached by identifying the processes that enable, facilitate, or initiate the development in question, whether or not it includes learning. Some a priori structure is always present to account for the particular sensitivities and responsivities of the organism at any point. But the a priori itself has a developmental history. The fact that learning seems not to play a part in the emergence of a behavior does not mean that particular conditions in a particular order are not required for its proper development. The fact that learning *is* important does not mean that species membership or developmental state is irrelevant.

Whether a characteristic persists in an individual's lifetime, on the other hand, depends on how it is regulated, whether the maintaining conditions are present, and on the way persistence is defined. Though proper conditions are often very reliable, both "innate" and "acquired" behavior will deteriorate in their absence, as will structure.

The presence of a character in all or most members of a species depends on the distribution of sufficient developmental interactants, nothing more or less. If these developmental systems show continuity across many generations, to that extent they are "phylogenetically derived"; they come to show such continuity, of course, by producing or bringing together those aspects of themselves that are not otherwise present. The history of such developmental systems and their divergences charts the linkages among near and distant relatives, and habitat is as basic to these mappings as are genes. Phylogeny is the derivational history of developmental systems. Opposing genetic to environmental factors as explanations for universals makes sense only under the theory of genes and environment as alternative sources of phenotypic form (rather than alternative sources of phenotypic variation). Similarly, the attempt to divide behavior that is variable in a species into that which is formed by conditional genetic instructions and that which is explicable by historical accident threatens to erect another three-quarter-way house on the nature-nurture course. Recall that Geertz's halfway house separated universals from nonuniversals. The next division separated universals into biological and cultural varieties. This third one would parcel out *non*universals into inner and outer causes, an admittedly neo-Zenoic strategy but an apparently tempting one. Once the questions themselves are separated and articulated, it becomes clear what sorts of things one would have to find out in order to answer them, and the irrelevance of concepts of genetic encoding of phenotypes becomes even clearer. What are presented as queries about kinds of development or about correct characterization of behavior are seen again and again to be questions about the variability and evolutionary history of organism-environment complexes.

Clearly, answering such questions requires detailed knowledge of developmental systems. Pronouncing traits to be biological rather than cultural answers nothing, but investigation of the relations between physiological and psychological levels might reveal a great deal about how those systems work.

If ontogeny is a matter of partially nested causal systems whose functional order is a result of the running of the system, not of any one set of its constituents, if phylogeny is progressive change in these developmental systems, and if, furthermore, the course of phylogeny is influenced by what can and cannot happen in ontogeny while ontogeny is constrained by many of the very functional requirements that shape the species history, then understanding of either requires "unpacking" the developmental system. Although a character may develop in any number of ways and still be adaptive, understanding how it evolved implies some knowledge of its development. Rather than being guided by some overarching strategy of identifying innate components or separating genetic and environmentally determined variance while considering gene-environment correlation and interaction an embarrassment (or simply excluding them from models), we might reasonably focus on the correlations and interactions themselves.

I have previously (Oyama, 1982) commented on a description of cross-fostering research in behavior genetics and on the problems that arise when one attempts to use determination of variation to explain development. Differences between inbred strains raised in similar conditions, McClearn and DeFries (1973, p. 167) say, are genetically determined. The between-strain differences may be reduced or eliminated by cross fostering. If this occurs, "then part of the initially observed difference between strains is ascribable to maternal behavior." What was previously genetically determined is now, in other words, environmentally determined. Genes are not thereby shown to be less important, say the authors, since the maternal behavior is still "a consequence of the mother's genotype," even though, to the offspring, such behavior is an environmental influence. Much that is a "consequence" of the mother's genotype in the natural world—including the entire developmental complex of prenatal and postnatal interactions with her, habitat, siblings and other conspecifics, other animate and inanimate sources of stimulation associated with her, and so on—also has developmentally significant effects. Such influences are environmental if they are varied experimentally, and genetic if they are not. What normally varies with the infant's genotype, that is, including many of its mother's characteristics and other features of its niche, can be experimentally detached from the normal system.

Cross fostering and similar research does not move us one step closer to an understanding of pure genetic effects by removing ever earlier and more intimate environmental effects, since the ultimate goal in such a program is the unintelligible one of arriving at naked gene "influences" in a vacuum, and since, as McClearn and DeFries imply, by eliminating such effects one eliminates most of the species-typical developmental system. Surely what is altered by such manipulations is often as much "a function of" genotype as what is left intact. What this research can do is unpack the developmental system by decorrelating variables that are naturally associated. The analysis in the preceding chapters shows that this does not isolate causes that act in the absence of environmental influences or that have their effects regardless of such influences. Nor does it expose the core of genetic reality in living beings. It may, however, show us about the timing of events and the susceptibility of processes to various kinds of perturbation.

If we view genotype-environment correlations and interactions not as methodological obstacles to be eliminated or circumvented but as a basic constitutive fact of development (King, 1968), we can set out to discover what is associated with what at which developmental stage and under what conditions, how the linkage is accomplished, what the adaptive implications of such development are, and what the consequences are of disrupting various associations. If patterns of variance are used as clues to developmental processes under various conditions, they do not have to be generally applicable to traits or species to be analytically useful, and the relativity of heritability becomes an aid, not a weakness.

It is not the ultimate a priori that is the Holy Grail here, but the nature of successive ones under various conditions. "A priori" in this sense simply means that which is "given" at any particular moment and which provides the organizing framework for subsequent interactions. In what ways does the emerging organism evoke, seek, produce, or reliably have supplied for it the very stimuli and conditions it requires for further development? How does it interact with them? What are the ways this entrainment can be derailed? What are the possibilities for compensation when such derailment occurs? Surely these possibilities are themselves dependent on conditions that may vary. How do some early variations become amplified, while others are damped? Propensities, potentials, stability, and lability are characteristics of systems in progress, and it is up to us, using the more productive lines of inquiry from behavior genetics, physiology, ethology, and related disciplines to analyze developmental systems of interest to discover how various inherited interactants are related.

When an environmental variable is found to be confounded with genotype, we have located an aspect of developmental order, an association that may well order future development. These are the associations we must investigate in order to understand how each structured developmental state arises, for it is this state that will be an important determinant of subsequent interactions. Such confoundings, then, are not dross to be eliminated by better experimental or statistical control. They may be the very gold we are seeking, and the point of control is to illuminate the ways mutual selectivity works.

Once the developmental system is analytically unpacked in a way that exposes not only the complex of organism-niche interactions but also the correlations among environmental variables that were described in the last chapter (between one that contributes to the development of a character and one that is adaptively important, for instance, as was the case for light and humidity in *Drosophila* emergence), we can see the manners in which such a system would succeed or fail in perpetuating itself. If all the interactants needed for the development of a character are present but are not associated with conditions to which that character is well suited (if dark-light transition occurs when humidity is very low), the system may not show transgenerational continuity. If, on the other hand, an organism is well suited to its living conditions but fails to assemble for its offspring the same developmental system that formed it, it may pass on its genes but its phenotype will not be reproduced. This is one scenario that accounts for "acquired" characters.

Both ontogeny and phylogeny thus depend on ecologically embedded developmental systems. Emphasis on such embeddedness is entirely consistent with the approach to natural selection described by Howard Gruber. In contrast to the sharp distinction between organism and environment found in narrower views of selection, he says, an interactionist description blurs that distinction in a number of ways, recognizing the importance of internal milieu as well. He notes that without coupling between organism and environment, neither adaptation nor the organism itself would be possible. If, however, coupling is extremely tight, the organism loses its identity (one thinks of organelles in cells, part of the "machinery" despite their distinct DNA). Loose coupling, he suggests, including coupling among subsystems, is important to the interactionist perspective (1983). It is helpful to realize, though, that loose coupling does not necessarily mean sloppy or unpredictable results. Recall Elsdale's inherently precise processes, or the out-of-gearishness that was so important to Raff and Kaufman's thinking, the redundant processes in which flexibility gives rise to exact consequences. A developmental process may be "rigid" in that it reliably produces similar outcomes despite perturbation; it can do so only by having a substantial amount of play in its workings. This is the point of canalization. Conversely, a process that is really rigid in its mechanics may fail to produce anything at all.

In addition to the image of the branching tree that is found in Darwin's thinking about evolution, an image so familiar to us and so readily interpreted as an icon of progression toward the human apex, there is another that is much less well known, one that appears near the end of *On the Origin of Species*. Gruber draws our attention to the "tangled bank," teeming with movement, variety, and interrelations, and suggests that this bank of connectivity provides the synchronic complexity that allows the "long-range diachronic process of evolution" to occur (1983, p. 12).

Gruber himself has attempted to synthesize the Great Man with historical processes, partly by looking at the multiple embeddings of "person-in-history, person-in-society, and person in a particular family setting" (Keegan and Gruber, 1983, p. 16). Though few of us are great persons, we are all part of our worlds. The Great Man issue is a special case of the more general one of the origin of ideas and events, and this in turn is a special case of the very general problem of the origin of form. As such, it invites a constructivist interactionist approach. In this chapter, I have moved repeatedly between discussing development and discussing ways of thinking about development. Though I have occasionally distinguished between data and their interpretation, there is no clear line between them, as each partially defines the other. As Gruber notes elsewhere, again on the function of images in Darwin's thought, "these images are not merely didactic or communicative devices; they seem to play a role in the actual generation of the theory: there is probably a complex and lively interaction between different levels of experience, such as the conceptual and the imaginal" (1977, p. 233).

We cannot clearly separate thought from its objects or the world from the way we construe it. In calling for an appreciation of the "plurality

of metaphors" found in Darwin's work, the "complexifying" as well as
the simplifying images, Gruber also advocates a less simplistically deterministic science, one that is more at home with weak theory, more sensitive to boundary conditions, a science that is interdisciplinary, inventive,
ample enough to embrace individuality, open, and humble (1977; see also
Klopfer, 1981).

We can think of programs for ontogeny because we can ourselves engineer programs and machines to execute them. When Descartes, whose
letter to More was cited in Chapter 5, argued that nature must produce
automata similar to the ones that man makes, he gave us a stunning example of the reasoning we have repeatedly encountered in our investigations: Since we do X when we imitate Nature, Nature must do X when
she does what we imitate. In the *Critique of Teleological Judgment,* Kant
spoke about our apparent inability (actually, he thought it was an absolute
inability) to think about nature without projecting teleological order into
it; he also spoke of the heuristic utility but final explanatory inadequacy
of such projection. Finding better images for development is presumably
difficult because the only way we know how to create self-creating life
is the same way less clever creatures do, a way that conceals the essential nature of its workings (as opposed to various details of its workings)
from our full understanding almost as effectively now as it did two hundred years ago. Perhaps it is no comfort that some people think that our
drive as builders derives partly from an obscure desire to mimic the creativity of nature. No comfort, that is, but the wan cognitive comfort that
comes from closing a mythic circle.

Most of my thoughts about prospects for our science and for ourselves as biological, psychological, and social creatures have been evident in the preceding pages, especially in the thematic reprise just completed. What follow are a few final observations on how we might direct our efforts in ways that are better fitted to the powers and limitations of science and that therefore better serve our needs as doers, knowers, and livers of lives in the real world. The nature of goals, which has proved so intractable to theory, will be addressed first. This will lead to a discussion of our need to reorganize our activities and expectations in ways that recognize our intimate interdependence with our surroundings. Finally I will return to the problem of what we are doing, and what we may hope to learn, when we read the universe as a system of signs.

A. Goals

If a major goal of much of our science should not be to distinguish under-lying, immutable biological nature from an overlay of nurture, and if workers in many fields have long known that this was in some sense the case and yet have had difficulty in consistently altering their thought and practice to reflect this knowledge, it is helpful to step back and ask why our questions keep framing themselves in such familiar ways. When we do so, we see there are many reasons, one of which is that we have set inappropriate goals for our science.

The ways in which order may be described; the ways it comes into being, is maintained, and inflected; its scheduling, effects, variations, and stability are indeed legitimate questions and are realistically addressed by analytic methods in all the fields we have mentioned. As we have seen, these issues are more easily treated by unpacking the developmental system to determine temporal priority, level of analysis, transfer of control, and interrelatedness of variables than by pursuing genetic or environmental sources of form or by opposing biological to psychological explanation in a mystifying matrix of nature-nurture ideas. Other problems, like the nature of existence, the constitution of the moral order, our fate as individuals and as a species, as well as certain notions of the normal, are associated with particular branches of science only insofar as our definition of these branches and their subject matter ensure that this will be so. (See Toulmin, 1982, pp. 21–85, on scientific cosmologies; he notes that physics is seen as "pessimistic" and biology as "optimistic." When biology is contrasted with the social sciences, *it* is in turn considered pessimistic.)

The nature of goals in the living world is related to the issue of what questions are scientific ones and what concerns are appropriately treated in a different vocabulary. This is because we have often construed ontogeny and phylogeny as goal-directed processes that set the immediate and long-range reasons and prospects for human life. What I have suggested is that goals are not to be inferred from regularity of constituents or outcomes, but discovered in kinds of causal relationships. While I have not attempted an exhaustive treatment of this problem, others have made important attempts in this direction, and more will do so, one hopes, in the future. Intentions and reasons, on the other hand, I am quite willing to treat as mental phenomena, whatever that may mean (Beer, 1983b), neither reducible to causes nor opposed to them or independent of them but describable in a different language, just as the psychological and physical are neither opposed nor identical, but exist on different levels of discourse (Rose, 1981; Toulmin, 1970). Goals do not require or explain minds and intentions, but characterize certain configurations of interconnected processes. To deal reasonably with them, then, we need postulate neither minds nor their mechanical surrogates. Nor need we deny, transcend, or augment "physical law," though we do need a concept of physical lawfulness that is more consistent with present-day science than with that of the seventeenth century.

A characteristic of certain systems in operation, not necessarily of a preexisting plan, a goal can change with circumstances. In ontogeny this is just what must occur for orderly developmental pathways, not just homeostatic stability, to occur. As effective variables and developmental state change, the set of causal interactions changes, and with it, the goals. Because we have tended to see life processes as intentional and therefore as proceeding according to preestablished plan, and because we have also regarded those plans as immutable, fixed by God or by a personified Nature, we have also tended to believe that our individual and species goals, or fates, were fixed. The distinction between intentions and contingent goals that are a function of real-life causal interactions, then, becomes much more than a philosophical nicety; it frees us from the complex of fallacies we have constructed around the concept of biological nature.

Whatever else biology may tell us, whether we are thinking of biology as the study of the physiological or as the study of the naturally selected (not, it is clear by now, the same thing!), it cannot tell us what is inevitable or, in any simple sense, what is desirable in our individual or communal lives. Some of the hotly contested candidates for "biological" status on the current scene may be more fixed (developmentally reliable under varying circumstances) than many hope, but less fixed than many believe. To know how fixed they actually are, we will have to do quite a different kind of research from the sort that has been assumed to be definitive on this question, a kind of research that, incidentally, does not identify experience with conditioning or confuse the relationship between biology and psychology with the relationship between cause and effect. Much of this research, furthermore, cannot be well-controlled, neat investigation, because it is usually human prospects that are at issue; it is much more likely to be social experimentation of the "let's try to make it better and see if it works" sort. How else can we learn the effects of conditions that have never existed before?

Because goals and other outcomes of mobile, interrelated processes are present neither in genes nor in environments nor in some grand evolutionary trend, knowing what is a necessary aspect of life depends on knowing where and how and by whom that life is being lived. In the real human world, at any rate, our dilemmas often involve the coordination of partially incompatible means and ends, of long-range, complicated planning with short-term strategies, of individual benefit with a common good that is often assumed to be opposed to it. Considering the gravity of some of the problems we face, it is incumbent on us not to prejudge these questions, especially on the sorts of assumptions we have looked at in these pages.

B. Putting Ourselves Back in the Tangled Bank

It is not the task of science, at least not the sciences that concern us here, to identify first causes. To be potent, predictable, and functionally effective, a variable need not be independent of other influences, even though we may vary it and call it an independent variable. It needs to be in the world in the proper way and can exert its proper effects only by being affected by the proper influences. Since its efficiency is not unlimited, boundary conditions need to be discovered, as well as its means of activity.

To shape an animal's behavior with conditioning methods, a trainer needs to be observant and sensitive to his subject. He needs to be under its precise control in order to control it. To control is not to stand outside the causal world; it is to rearrange oneself in it. This is true when we do science, and it is true when we just do life. We don't really have to reinsert our science or ourselves into the tangled bank, because we never left it. We do need, I think, to recognize more fully and explicitly the extent of our entanglement; it is the source of our power and of our vulnerability, and it is easy to be blinded to the latter by our love affair with the former.

In our scientific analyses we bow to causal embeddedness in tacit ways, even when our language ignores or denies it. (Recall the descriptions cited in Chapter 3 of environmental contributions to ontogeny and the convoluted attempts to reserve formal or essential influence for the genes.) We do the same in civilian life. An executive who wishes to control a market must be responsive to a multitude of complicated indices. The more ambitious she is, the more things must be taken into account (or, and this is more risky, assumed), and, in a way, the more exposed she is. A farmer increases his crops by using certain chemicals. He thereby comes under the immediate and long-range influence of a large set of economic, technological, and ecological factors. Eventually he often degrades the very land and water he is controlling, improving himself right out of business. Attempts to control without sufficient regard for interdependence in international relations are equally ill-fated.

What we call mature, independent conduct is not, in fact, an ability to affect the world without regard for the world's needs and demands. A fully socialized being has become a certain kind of participant in the larger social network; "independence," childhood fantasies notwithstanding, really means interdependence with the extended adult world, not freedom from requirements and influences. It is quite disturbing to encounter someone who seems to be beyond influence. Even the insane, however, are often more aware and "tuned in" than they appear.

I do not suggest that all kinds of control are psychologically, ethically, technologically, or politically equivalent. I do contend that to view control as one-way flows of information or influence is to miss half the story. The special talent that emerged with the evolution of humans is not for independence from the environment or the simple control of it. It is rather for extraordinarily subtle, multilayered analytic abilities that include the capacity to contemplate the far past and the distant future, the immediate and the remote environment, and multiple possibilities in each. With this ability we spin webs of interrelatedness unprecedented in their variation, complexity, and ramification. A possibly mythical Murphy said that anything that can go wrong, will. While we may dismiss him as a crank, what we are often doing as we extend our control is multiplying the things that can go wrong; all these things, in turn, are connected in ways that defy even our analytic ability.

If we think that our evolutionary position makes us lords of the universe, we might remember the difficulty we have long had in deciding whether a God who makes the universe and its laws once and who is thereafter removed from its workings is consistent with our vision of a loving father who sees us, hears us, and intervenes on our behalf. On the other hand, can a God who is moved by his children's plight and who answers prayers be an unmoved mover? Gods who meddle too much in the world become as subject to its pleasures and its pains as mortals. In the Greek pantheon, it is sometimes difficult to separate the naturals from the supernaturals.

C. Reading the Universe

We have alluded many times to the ways in which the vocabulary, motives, and practice of contemporary science are continuous with, or at least resemble, those of the devout decipherers of God's will and nature who gave our science birth. Toulmin (1982, pp. 21–85) observes that to call a theory a scientific myth is not to dismiss the theory, but rather to say that it is persistently used to answer nonscientific questions. Though our forerunners were ignorant of many things, he also points out, they did not invent myths to explain everything. Atlas reassured them by holding up the world because they felt some degree of insecurity about the world and all that its stability/instability symbolized. One of the problems we have encountered in treatments of development is the counterfeit comfort that accompanies the description of a mysterious process in mechanical or information-control terms. It may indeed be comforting (or discomfiting, depending on one's beliefs about the content of genetic programs) to believe that our fundamental nature as individuals or as a species is set before we see the light of day, and that the processes that get us that far and beyond are guided by some transcendent intelligence accumulated over millions of years.

What one reads in the genes (by surveying related species as one would texts by authors of the same school, by comparing phenotypes of a single species, by failing to perturb the development of some character, by watching a behavior appear without learning, by deciding that a trait contributes to fitness, etc., etc.) is not related in any simple way to what one wishes or believes about human prospects. That relation depends on our concepts of will and possibility. Some call on us to recognize the genetic message and to combat it, to defy it, to make it false. In this case biological truth is used as a cautionary myth. Cartmill, for example, feels that views of humans as a threat to the balance of nature are accurate, but that they also express a vision of nature as sacred, a vision that may be useful in preserving nature (1983). The image of man as destructive beast, then, is not necessarily part of a worldview that endorses or even accepts destructiveness. It does, however, convey important things about our self concept and our values. In an interesting aside, Cartmill suggests that this mythic role of science can limit inquiry: focus on the brain as primary to human evolution, clearly a function of our idea of ourselves as intellectual kings of the phylogenetic mountain, led scholars to overlook the small australopithecine skull as an important link in our descent. At a much cruder level, of course, one finds (still) protestations that evolution can't be valid because this would make us "nothing but" apes. (What is crucial here, of course, is not the "apes" but the "nothing but.")

We tend to roam the ethological and ethnographic literature for signs indicating our nature and our future. Whether the genealogies we draw are self-congratulatory, darkly fatalistic, or, in Shklar's (1971, pp. 137, 147–148) terms, subversive exercises in "reductive unmasking" with or without programs for change, they refer, as she points out, to a past that is present in the descendants. This neatly describes the role of the gene in conventional thought, though Shklar is not speaking of the phylogenetic past. Such mythmaking is "neither pseudo-history nor pseudo-etiology nor primitive science . . . neither the rival nor the precursor of more rigorous forms of thought. It is psychological evocation." It arises at times of great confusion, when understanding and resistance fail. If we reject original sin and are basically good, why are our lives full of pain? The need to blame, to attribute responsibility for suffering and injustice, is one of the things that motivates myth.

We have tried to educate and aid the poor, and the poor are still with us. We have claimed our legacy as lords of the earth, and the earth is disintegrating around us. We have tried to construct an egalitarian state, but gender, class, and ethnic boundaries still vex us. We "rehabilitate" criminals, and they victimize us again. We modernize underdeveloped nations, and they revile us. We construct a therapeutic ethic of individualistic freedom and fulfillment and find ourselves adrift in an ever more sophisticated wasteland. We manipulate the atom's nucleus, and it threatens to annihilate us. Even our genealogies confound us: we claim the peace-loving chimpanzees as our nearest relatives and they subvert our moral phylogeny by killing each other.

White (1972, pp. 239–242) suggests that the historical past is "plastic," as we choose our ancestors to justify our changing definitions of ourselves. "Social fatherhood," he claims, is "bestowed by the sons." The end of the Roman social-cultural system came when people stopped regarding themselves as descendants of Romans and started thinking of themselves as children of Judaeo-Christian forebears instead. To choose a past is thus to choose a present, and vice versa. "By constructing our present, we assert our freedom; by seeking retroactive justification for it in our past, we silently strip ourselves of the freedom that has allowed us to become what we are."

White contrasts this retroactive construction of history with the fixity of biological ancestry, but I think that we range across the evolutionary landscape in much the same way that we select our forerunners from the more recent past. We use fossils and contemporary similarities and dissimilarities among species to discover phylogenetic history; we also use ontogenetic and phylogenetic history to discover ourselves. I would not be so brash as to speak out against the mythic, cosmological impulse. To some extent, perhaps I am engaged in mythmaking myself. I do think, though, that we ought to recognize the impulse and try not to mistake it for "simple" fact gathering, and that we ought to take as unflinching a look as we can manage at those aspects of our lives that give rise to these anxious divinings. If we are truly concerned with dealing humanely and realistically with ourselves and the world, then we cannot afford prematurely foreclosed possibilities or naive, simplistic optimism, crossed or circular inferences, empty explanation, or facile analogy. We cannot let projections (of ourselves into our genes, of our past into our future) pass for understanding, or disciplinary ambition for theory. I am suggesting, that is, that some worldviews are better suited than others to our abilities, including our ability to do science, and to our requirements as denizens of the natural world.

If we succeed in enriching and broadening our ideas of causality, in placing information in the context of the processes that produce and reveal it, in grasping the ways variety supports uniformity and vice versa, and in placing ourselves back in the tangled bank that is our home (or as Toulmin puts it in a discussion of the separation of mind from matter, reason from emotion, humanity from nature that has characterized the modern pursuit of detached knowledge, in reinserting humanity into nature, 1982, p. 262), we will have accomplished a great deal. And in the bargain we will have done more justice to the genes, to the world that houses them, and to our own constructed nature than we have in the past.

In his story "The God's Script," Jorge Luis Borges, master of multiple embeddings, infinite branchings, and ambiguous boundaries, tells of a god who, foreseeing devastation at the end of time, wrote at its beginning a sentence capable of preventing that destruction.[1] This sentence, designed to resist time and chance, becomes the object of the thoughts of an imprisoned man. The captive realizes that even in human language, each statement implies the universe:

> To say *the tiger* is to say the tigers that begot it, the deer and turtles devoured by it, the grass on which the deer fed, the earth that was mother to the grass, the heaven that gave birth to the earth. I considered that in

the language of a god every word would enunciate that infinite concate-
nation of facts, and not in an implicit but in an explicit manner, and not
progressively but instantaneously.

The man deciphers the message but never utters it, having finally
grasped the wholeness and interconnectedness of the universe, a unity
that undoes individuality. "May the mystery lettered on the tigers die with
me. Whoever has seen the universe, whoever has beheld the fiery designs
of the universe, cannot think in terms of one man, of that man's trivial
fortunes or misfortunes, though he be that very man."

Can it be that if we really reinsert ourselves into the world, see our
development, investigations, and technological control as actions within
a network that we support and alter and that supports and alters us, see
freedom and responsibility not as denials of causality but as a particularly human acknowledgment of it, if we see nature, including our own,
as multilayered and constructed in development, not prior to it, if we see
the world as truly our home and not, as Toulmin acidly remarks, our hotel
(1982, p. 272), with all the loving reliance, multiple attachments, pride,
and farsighted maintenance that "home" entails, is it possible that we will
no longer need a mystical hidden message? Is it possible that the only
message is our lives in our world and the life of our world in its universe?

Afterword to Second Edition

A great deal has happened on the intellectual scene since this book was written. Nonetheless, I discover that I agree with myself quite thoroughly, so these concluding comments include no dramatic recantations. Informed as it is by hindsight, an afterword does allow for second thoughts, judicious reiterations, and clarifications, and I'll indulge in a few of each as I reflect on what has and hasn't come to pass in the last decade and a half. I begin with a pair of recent variations on the theme of information and some remarks on interaction. Then I take up a pair of issues that come before me with some regularity. One is what viewers of U.S. television commercials will recognize as the "Where's the beef?" question, or (in more information-theoretic terms) "What difference does this difference make?" The other has to do with defining the boundaries of a developmental system. I close with some directions in which I would like to see the developmental systems perspective further developed.

A. More Variations: Information and Interaction

The Ontogeny of Information was written in the early 1980s, amidst the furor engendered by sociobiology and other developments in biology. Then, as now, there was a great deal of highly charged language and ambitious theorizing, and since then the varieties of information talk, with or without a genetic connection, have gone on mutating and proliferating. As the linked technologies of molecular biology and computer science have expanded, infospeak has become even more widespread and deeply embedded in the culture.

In Chapter 1, the *cognitive-causal gene* was introduced to refer to the descriptive conventions that had molecules and cells recognizing, selecting, and instructing each other. I do not use this particular term anymore, but its offense was gracelessness, not inaccuracy; indeed, its flat-footedness can focus attention on the confused state of our discourses about genes. To some extent the confusion is traceable to the usual suspects, such as carelessness, gaps between technical and lay usage, or overlapping, ambiguous, and unstable terminology. There are, in addition, the theistic overtones I've discussed, which have been noticed by other commentators as well.[1] Writing in the *Atlantic Monthly,* Cullen Murphy (1997) remarks on the "nucleotidal wave" that surges around us, and I have even seen reports that entrepreneurial souls are hawking trinkets guaranteed to contain the DNA of the stars. This is surely the (not-so-) secular successor we deserve to the reverent preservation of the jawbones and toes of saints—if you will, our nucleotidolatry.

I. MORE ON INFORMATION: DISHRAGS AND MINDS

Two works of philosophy, both aimed at fairly broad audiences, show how further variations continue to be wrung from the information theme. Each in its way also demonstrates the pitfalls attending any attempt to make information do serious theoretical work.

The first is *Darwin's Dangerous Idea* (1995; all references are to pp. 115, 196–198), Daniel Dennett's defense of his selectionist view of life and mind. In a chapter on biology as engineering, he answers unnamed "deconstructionist" critics of " 'gene centrism' . . . the doctrine that the DNA is the sole information store for inheritance." There is more to gene-centrism, but the focus on genes as the carriers of inheritance is certainly a central element, as is the resulting necessity of deriving organisms from those genes. The language of selfish, calculating genetic agents that Dennett sometimes adopts undoubtedly draws complaints as well. Here, though, I wish to direct attention to what remains unclear in these passages: just what it means to inherit information.

The critics, Dennett reports, deny that information is carried exclusively in texts: Some of it must be brought to the text by the reader. These deconstructionists, he says, are correct, both about literal texts and about the biological "text" of the DNA, which needs readers of its own in order to function as information. I've argued in this book that the supposed autonomy of macromolecular representations buttresses claims for the genes' special powers. By admitting the informational insufficiency of

DNA, Dennett appears to give up a great deal, though because he denies that he ever accepted the doctrine of complete information to begin with, he doesn't present the admission as a concession. In fact, he insists that the idea of inherited information being carried completely by the genes "was always only a handy oversimplification." Then he blandly allows that, just as a library contains no information without additional information from readers (who carry it mainly in their DNA), DNA also needs readers, and that "the language of DNA and the 'readers' of that language have to evolve together; neither can work on its own." He does not say what is included in the "code-reading environment" that supposedly contains the remaining information needed to make an organism. As we shall see below, this environment is not just the rest of the cell, which is where most people draw the line when they extend inheritance beyond the chromosomal DNA. One wonders where he does draw the line, or whether he is happy to have information *everywhere,* and if the latter, what the consequences are for his larger project.

Dennett declares that the materials and conditions provided by the environment complete the information in the DNA. Recall that describing the environment in terms of materials and conditions is a classic way of minimizing its formative role. Here Dennett seems to be trying to do just that, and at the same time, somehow, to have his information be there, too. The moisture that allows bacteria to live in a dishrag, he says, is "part of the informational background presupposed by the DNA of the bacteria." We therefore dry rags between uses. He continues, "any *functioning* structure carries *implicit* information about the environment in which its function works." This is reminiscent of the attempts, reviewed in Chapter 5, to deal with environmental relationships while retaining directive power for the genes, by invoking open programs, conditional rules, and the like. We saw that trying to handle such contingencies by putting more (here perhaps only "implicit") information in the genes fails because it commits the theorist to one of two dubious strategies: positing foreknowledge of *all possible* causal encounters or only of *adaptive* outcomes. Dennett seems to be attempting the second. But recall that this second tack, of encoding certain outcomes but not others, makes it difficult to explain the occurrence of events that are not foreseen by the genes. If some biological interactions can take place without genetic recipes, what distinguishes the ones that need the extra help? If bacteria die without moisture, furthermore, it is difficult to see why information "about" moisture, implicit or explicit, is needed to explain all those little demises in Dennett's kitchen.

Dennett starts a discussion of function and specification by asserting that Monod (1971) "quite conclusively solved" the problem of inadequate information in the linear sequence of the DNA molecule. In Chapter 3 I discussed Monod's emphatic assertion that only the genes had the power to specify their products. For him, the cellular environment contributed to the three-dimensional shape, but had no *specifying* role; protein function was completely defined by the genes. This was one of the linchpins of Monod's cosmology, including his view of development as revelation, and it helps to anchor much genecentric thought. Yet Dennett wants to use Monod's "solution" to show the precise opposite: that the genes alone cannot specify anything, that the task of specification, which Dennett calls *"the function on which all other functions depend,"* must instead be distributed.

It is not clear whether Dennett senses that his entire concept of information is in peril. He seems to waver between decreasing and increasing the informational load on the DNA. He does not acknowledge the contradiction between his position and Monod's, instead hailing the French scientist's formulation as a resolution. Nor does he seem inclined to follow the implications of his own statements. Suppose we take seriously his disclaimer that the "information store for inheritance" was never *really* meant to be confined to the genes. Was it really meant to be in damp dishrags? As I pointed out above, this is not just a minimal enlargement of hereditary transmission to include other cell constituents; it reaches outside the cell itself. Once you have scattered inherited information around with the very free hand that would be needed to take care of all biological functions, what happens to the traditionally privileged channels of heredity (genetic and, in dual-inheritance models like his, genetic and cultural)? This is a slippery slope, as we shall see below. I have quite happily slid down it, but it is a trip one should take with one's eyes wide open, and one should not expect information to look the same from the bottom. Perhaps Dennett means that information is not "out there," that it is not in the nucleus or anyplace else, that it is a way of talking about certain interactions rather than their cause or a prescription for them. If so, it cannot be carried, stored, or transmitted at all. But then Dennett would seem to have ended up in agreement with at least some of the critics he finds so obtuse.

One only has to strain to reconnect genes to their environments (which are not macroenvironments like rags and libraries, but the small-scaled ones of other molecules), and to reconnect organisms to *theirs* if one has analytically detached them in the first place and then undertaken to explain their doings by means of abstract representations rather than the same sorts of everyday material interactions by which the rest of the world runs. One way or another, Dennett has conceded not just that environmental "continuities" are indispensable, but that the whole idea of genetic information is senseless without them, and that one cannot consistently withhold informational status from the aspects of the environment that make the genetic information informational. This was a concession Monod could not grant. Dennett appears to be ambivalent about where to locate the informational complement to his bacterial genes, but persuasive momentum seems to mean more to him at this juncture than rigor. His heavy reliance on the boundary breachings of Dawkins's (1982) elastic phenotypes, which I discussed in Chapter 8, positions him to make the move that renders the rest unnecessary: to put the genes back into the interactively constituted system in which they are no longer informationally privileged prime movers — where, in fact, one cannot speak of *stores* of information at all, and where relations with the surround can be given proper weight without the conceptual contortions they otherwise require. But this is not what he and Dawkins are about, and he seems no more willing to make that move than Monod was to renounce *his* version of genecentrism.

The flight into informational space is in many respects a flight from materiality. For those committed to a functionalist framework that detaches the patterns that *really* count from their material base, this is understandable. But, as I have suggested, it is ironic in the light of the aggressively materialist, scientific stance frequently adopted by some of its practitioners, and it can leave them at a loss to say how things actually happen.[2] Interposing a layer of information and computation between cognizing entities and their surroundings does not seem the optimal solution.

A more audacious effort in many ways is David Chalmers's *The Conscious Mind: In Search of a Fundamental Theory* (1996; all references are to this volume). Not as close as Dennett's to the biological issues that are the focus of *The Ontogeny of Information,* this book nevertheless also makes heavy use of information concepts.

There are two general ways to use mentalistic language for things other than ourselves. The restrictive one I have favored is to say that such locutions are fine for ourselves and, perhaps, certain other animals, but are not to be taken seriously when applied, say, to molecules. To be sure, it is not a simple matter to say precisely what we should mean when we talk about ourselves in these ways, but *whatever* we mean, it is different from what we mean when we talk about DNA. The expansive route, taken by Chalmers, applies mental language more generally by associating it with something else, in this case with information.

Like me, Chalmers regards information as a reducer of uncertainty, a difference that makes a difference. He seems inclined, however, to treat it not as observer- and context-dependent, but as simply *there.* He suggests that information, and so experience, could be everywhere there is causal interaction. In his eighth chapter Chalmers unapologetically lays out the implications of identifying mind with "information spaces," and though he calls his discussion speculative, he does not flinch at the panpsychism that can come of locating information everywhere in the world. Well, perhaps he flinches a little: A rock may not have experience or consciousness, he suggests, because it isn't "picked out as an information-processing system" (p. 297). I would ask, "Not picked out by whom?" A geologist reading striations and stress lines could regard a rock as an "information-processing system," though it is not clear what would be gained by doing so. Rocks behave in regular ways and present discernible differences that are correlated with other differences in which the geologist may also be interested. They can transform continuous changes into discontinuous ones, diminish temperature swings, or produce other feats of "processing." But it is the geologist analyzing the rock, or the architect using stone as a heat reservoir, for whom I would say there is information. To explain the rock's structure and relations by information is to *entify* what is better treated as an epistemological notion.

Though I have some sympathy with Chalmers's obdurate insistence on including experience in our accounts of the mind, I doubt that information processing is the best platform from which to launch an investigation into such treacherous territory. In addition, a possible embarrassment of experiential riches is engendered by his tactic of locating mind wherever there is causation, something he acknowledges himself (pp. 300, 303). "Experience is information from the inside; physics is information from the outside," he declares. He also seems to realize that there is no apparent limit to the number of differences that could be generated by drawing different distinctions. Chalmers does not pursue this point, but different differences can make a difference, depending on what question is being asked, and by whom. So when I speak of information being generated in causal interactions, there is a perspective from which this is the case. The complexity—the differences—that we observe in an embyro, say, cannot be entirely accounted for by the complexity we see in the DNA, and when I say a difference makes a difference for the embryo, there is someone picking out just these differences. This does not mean an observer is needed for things to happen at all, just that information talk adds nothing to whatever descriptions or causal accounts one wants to give. I suspect that stones break and trees fall in my absence, but the difference-based definition Chalmers and I share would seem to bind information inextricably to a point of view. Yet he seems to want something close to the precise opposite: to derive point of view—experience, consciousness—from information.

These two treatments supplement the set of variations in Chapter 5. One could take them as evidence of the futility of trying to stem the informational tide, nucleo- or otherwise. It's just possible, however, that by approaching the limits of their schemes, these authors are inadvertently contributing to the critique. Dennett works against his larger aims with the concessions (that genes cannot specify anything on their own, for instance) and expedients (information in readers and rags) he needs to make his fairly colloquial treatment seem plausible. Chalmers shows that if he tries to treat information more rigorously, there is no way of getting it to play the role he wants it to play in the generation of mind.

2. INTERACTION: PARITY AND SYSTEMS

In the preceding section I suggested that one can take a restrictive or expansive attitude toward mentalistic terminology. In certain respects, though, which choice is made is less important than the way it is put into practice. My principal quarrel with those who speak of genes that foresee, recognize, or organize is not so much that figurative language is used for molecules as that it is used for some molecules and not others, and not only to enliven a narrative, but to *explain* how the living world works. I have maintained that these asymmetrical applications create more difficulties than they solve, perhaps even feeding a subtle biological determinism that the analysts themselves may try to disavow.[3] If a molecule can "recognize" appropriate objects because it interacts selectively, then other molecules that interact selectively (what molecule does not?) should, in the absence of reasons to do otherwise, be described in this way. But then the cognitive powers of the first bit of matter cannot be used to elevate it above the second one.

Genetic explanations, both popular and academic, have frequently relied on some version of the argument from the poverty of the stimulus (POS): the organism displays too much structure, too much knowledge, too much skill for experience or learning to be an adequate explanation. But I have argued for a corresponding "poverty of the gene," a POG to the nativist's POS, and have maintained that, far from minimizing the significance and complexity of DNA interactions, this parity of poverties places those complexities, and countless others, squarely before our noses.

Not mentioned before this point, the philosophers' phrase "parity of reasoning" is another way of referring to the search for logical consistency and intellectual fairness that drives so many of my arguments. Symmetrical application of a piece of reasoning has an ability to "push" for specific justification, unearthing hidden assumptions and covert privileging. One will then have to muster persuasive reasons specific to the processes in question in order to argue that some causes are more equal than others. If my analysis in the previous section is correct, a conceptual scheme is being pushed, and the reactions are themselves informative.

An interactionist impulse toward inclusion and parity can be indulged in more or less restricted ways. One may include both biology and culture in one's scheme and make the obligatory antigenetic determinist noises, but to go on funneling them through different channels of causal influence and inheritance is to lack the full courage of one's interactionist convictions. In Dennett we see a juxtaposition between substantial acknowledgment of informational insufficiency, interdependence, and coupled change, on the one hand, and quite traditional dual-channel notions of heredity, on the other. If parity of reasoning is joined to constructivist interaction and turned on the concept of heredity itself, it leads to something like a developmental system.

The difficulties with the notion of interaction I alluded to in the book have not disappeared. It may well be that the scrupulous addition of "constructivist" in my more recent writings and in this revised edition will still not be sufficient to head off conservative readings (conservative because they retain some form of the developmental dualism that the concept of the developmental system is intended to replace).[4] It would certainly be unfortunate if I have contributed to the very misreadings that dismay me now, by continuing to use "interaction" or by writing about genes and environments (or genes, phenotypes, and environments) evolving together. What I meant to communicate was the unwisdom of treating either independently of the other(s), the impossibility of having predefined natures or environments at all. The trade-off between maximal comprehensibility to readers and fidelity to one's own vision is a delicate one, and the fact that scholars who use different terms are similarly misunderstood suggests there is no easy solution (Oyama, in press-b).

Asking about parity in developmental causation doesn't just help us discover how categorizations and privilegings are justified; it leads to conceptual reconstruction if the justifications don't work (see Oyama, 1998). I still think that the notion of system can be useful here, though its connotations are more divergent and troublesome than I realized in the past. For some, as I indicated then, calling a causally interconnected assemblage of factors a "system" marks it as highly reliable and resilient: It becomes a black-boxable mechanism, as when someone claims to have a system for picking winning horses or stocks. For others, to invoke a system is to emphasize the *un*predictability, unintended consequences, even practical uncontrollability—of all those contingent interactions. (For a variety of views, see Jervis, 1997; Rosenthal, 1984; Taylor and García-Barrios, 1995; Vayda, 1996.) For me, what is salient about systems is the way that connectivity does away with in-principle and cleanly distinguishable causes and effects: the dependence of causal relevance, process, and outcome on changing configurations of factors. I have tried to avoid prejudging degree of regularity. Over both developmental and evolutionary time, the interesting questions are often precisely about when, how, and under what conditions transitions occur between reliability and unreliability of a process or product.

B. Where's the Beef?

Much of this book is devoted to analysis and conceptual overhaul. A critical project that is on target is worthwhile in itself. By challenging existing practices it can lead to a sharpening of researchers' goals, methods, and conclusions: they must state more precisely *what they are doing and why they are doing it.* In addition to the critiques, or rather, simultaneously with them, I have presented an alternative conceptual perspective on development, evolution, and the relations between them. This alternative yields new ways of interpreting and elaborating on observations, and suggests other lines of research. Along the way I have cited a great deal of relevant empirical work from a wide variety of fields. Yet I regularly encounter the "Where's the beef?" question. Sometimes it amounts to a pugnacious poke in the chest, sometimes, perhaps, a check on just how different the vision really is, or a probe into the possible costs of change—of shifting direction or reinterpreting findings. The demand can even be an attempt to avoid confronting the gaps in the questioner's own framework.

To show what difference this alteration of perspective can make, I first take up the levels at which a research program can be affected. Next I turn to the opening up of investigative possibilities that can occur when a developmental systems perspective is adopted, and finally I examine the benefits of reframing.

I. CHALLENGING ON MULTIPLE LEVELS

The challenges issuing from the developmental systems perspective can be taken on at more than one level. Sometimes a reference to a program or similar entity seems to involve an empirical claim. Once we see that there is no single referent for "program" or "nature," however, we must ask precisely what the claim is, and we usually find ambiguities even at the level of conventional understandings. In some cases the research implications of a conceptual readjustment are absurdly obvious, though the resulting work is apt to be of a quite familiar sort. Universality within a species, developmental stability in individuals, and the absence of a certain kind of learning history, for example, are often thought to go together because in the traditional framework they are all signs of "genetic control." But they are not necessarily associated, so each must be demonstrated on its own. Blocking the easy but reckless path of *assuming* along nature-nurture lines opens up worlds of (researchable) uncertainty.

If it is important to ask researchers to say clearly what their research does and does not show, it is equally important to catch the slide in logic, the unspoken and perhaps untenable assumption that gives some findings a significance they do not merit, or too hastily denies significance to others—to point out that two findings are compatible even though the terms in which they are couched suggest that they are not, or to show that two issues are related even though they are treated as though they were independent. As we saw in the section on interactionism, the concept of the developmental system is sometimes construed in quite circumscribed ways (biology and culture are both important, etc.). I suggested that the long-term value of the approach is related to the extent of the theoretical realignment. Merely substituting terms (saying "species specific" or "canalized" instead of "innate," but *meaning* "innate") is worse than useless because it implies a change where there is none. If scientists warn their readers that a "genetic basis" does not necessarily signal inevitability, a modest gain has been made, though which meaning of "genetic basis" is intended must be specified, for the phrase participates in pretty much the full range of ambiguities detailed in the preceding pages.[5] Similarly, indicating that biological and cultural factors influence each other can strengthen a treatment in some ways without touching the more basic two-channel model. (For an alternative approach to human evolution, see Caporael, 1997).

Consider the kind of work in which family trees and gene mapping are used to find genes "for" diseases. Time after time they are found, and time after time, but with less fanfare, claims are moderated or taken back. If one goes beyond the headlines and lead stories on the evening news, one usually finds that the scientists gave the proper caveats: ". . . only this sample . . . disease is complex . . . celebration premature . . . more research needed." This pattern doesn't necessarily show that the research itself was fundamentally flawed: maybe there *was* a substantial relation between some genetic variant and the disorder in that population, say, and that relation might have practical import, the scope and nature of which are not yet clear. Failure to attain total generality is not total failure: such results can and should be exploited for research leads. To take a knee-jerk oppositional stance ("See, *of course* genes have nothing to do with it") or simply to sigh hopefully ("Maybe they'll find the real gene *next* time") is to miss an opportunity to inquire into the conditions under which relations do or do not appear, to say nothing of the whole way genetic influences are understood. Not only can chromosomal segments appear with different frequencies in different groups of people, but their involvement in pathological processes may well depend on factors that are variably associated with different populations as well.

K. C. Smith (1998) gives an example: Suppose a gene "for" breast cancer is found in a particular population. Women may be told that they have an 85 percent risk of developing certain cancers, even though they may belong to a different population and may thus be subject to a different set of factors. If so, their statistical risk may well be different as well. It can be very difficult to track down the other factors and their interrelations; this is why caveats are necessary. While researchers are certainly aware of such contingencies, and often include them among their lists of cautionary hedges, they may also indulge in common forms of gene talk (so common that they barely qualify as such) that undercut caution. Saying the gene "carries" or "confers" risk obscures the possible context-variability of the DNA-disease links, implying that risk simply travels with the molecule.

The perspective developed in this book also raises more fundamental questions about the whole enterprise of hunting for magic genes that code for particular phenotypic effects.[6] Things can get more difficult when one tries to move beyond the usual techniques, and geneticists may candidly admit that they look for genes because it is easier than studying other factors. An altered research strategy might reduce the boom-and-crash cycles, which often benefit the purveyors of hot medical news more than they do the researchers who produce the results or the public at large.

My case against the genetic program and kindred constructs, then, shouldn't be mistaken for an empirical claim of the usual sort, where the only thing that is uncertain is whether or not a particular outcome occurs. It is the intelligibility, utility, and unity of the idea of a program that I contest, and this is different from saying, "You and I both know what it is, and it could be in there, but I've looked, and that space is vacant." At the same time that the developmental systems perspective blocks a variety of assumptions and interpretations, it raises more vexing, more general, and, to me, more interesting problems: Why were these features assumed to go together in the first place? If the notions of biological reality and control that support the programming complex are shown to be defective, what is the theoretical motivation for pursuing just this set of questions?

2. OPENING OUT

In Chapter 3, I pointed out that a gene "initiates" a sequence of events only if the investigation starts at that point. Certainly, one can start there, but the challenge to the scientist is to own up to posing the question this way, and thereby limiting the possible answers, and perhaps to try other, less immediately available, questions as well. Working from an idea of an interactive complex, a researcher might be more apt to bring the taken-for-granted background to the foreground. Recall the utility of parity questions in increasing the depth of the cognitive field. Especially when used for specieswide patterns, the notion of genetically controlled development doesn't offer much in the way of research leads. Usually it just points to the predictable outcomes that so often allow us to take development for granted, whereas asking how control emerges and is transferred during development can lead to work on transitions and changing modes of regulation (Deacon, 1997; Thelen and Smith, 1994; Turkewitz and Devenny, 1993). This in turn can extend the investigative frame beyond the usual limits, for genes, organs, and organisms do not necessarily respect the same boundaries we do.

When I say that much of the developmental complex is passed on in reproduction because its constituents are tied to the organism's life cycle, I am indicating alternative conceptualizations of development and evolution and their research programs at the same time. The breeding failures mentioned at the end of Chapter 3 are but one kind of practical impetus to searching beyond the obvious prerequisites for reproduction; conservationists' tales of endangered developmental systems are legion. More to the point, such failures raise the question of what, exactly, must be re-produced for a new generation to arise, and how this comes about. A changing set of relations must be reestablished, maintained, altered, and terminated, even when the interacting entities need not be literally re-created in each generation (as does a placenta, for instance, and eventually the maternal body in which placentas are made, or, if it is used only once, the nest or burrow that shelters that body).

As suggested in the cancer genetics example, the nuisance terms generated in research sometimes bear a second look. Those statistical interactions and correlations between genotypes and environments may reveal more than limitations on the internal and external validity of the findings; they may also point toward some of the very connections we seek. The studies of extended networks cited in this book are relevant for such a project: the coevolution of prey and predators, symbionts and parasites, including much of the work drawn upon by Dawkins (1982) for his extended phenotypes. Once scientists are equipped with an enriched sense of developmental and evolutionary resources and processes, their research areas will widen.

As the notion of centralized control of development gives way to the developmental system, the view of heredity is enlarged beyond the germ cell to encompass other developmental means or resources. As indicated above, this increases the number of ways in which developmental influences can act transgenerationally, altering the relationship between developmental and evolutionary processes. Think of the way a play can sometimes be "opened out" when it is translated to film, showing things the stage couldn't accommodate and moving the story forward or adding dimension to it. For some scientists, the empirical extensions I am suggesting require little or no redirection. There is a great deal of work now being done that fits nicely within a broadened developmental systems framework, which is not tied to particular modeling techniques, for instance, as connectionism or dynamical systems theory is. Other researchers might have to do more realignment, but they won't have to do it by themselves.

A nice example of the conceptual and empirical fruitfulness of the "pushing" function of parity analysis is found in an exchange between Sterelny, Smith, and Dickison (1996) and developmental systems theorists Griffiths and Gray (1994, 1997; Gray, 1992). Sterelny et al. recognize the nature of the conceptual challenge and try to *work* with it (rather than waving it away, as Dennett appears to). They conjecture about a variety of "routes across the generations" by which variations in animal burrows might be reconstructed, perhaps by affecting their tenants' development.[7] Such developmentally and evolutionarily significant interactions can also take place among plants or between plants and animals. These theoretical efforts simultaneously produce engaging analyses and research ideas by making it possible for questions that have been hitherto pursued asymmetrically to be investigated with respect to other (by hypothesis, comparable) entities.

Gordon (1991) has written about the life cycles of ant colonies, which change on a longer time scale than the life spans of the ants themselves. How are these cycles linked, and how are they related to larger ecological stabilities and changes? Such questions are natural companions to those about cells, with their rapid turnover and small size, and the organs or organisms they help to make and in which they are shaped and sustained.

It is important to realize that these intergenerational routes can't be slotted cleanly into the usual gene-culture coevolution schemes. The nature-nurture dichotomy is variously expressed as genes versus environment or experience, biology versus learning or culture, and so on. But slicing on a different angle doesn't freshen stale bread. "Channels" are finally attempts to circumvent the processes of developmental construction. One could say the research program under discussion is nothing less than the elucidation of the constitution and dynamics of an indefinite number of developmental systems.

3. REFRAMING

Elsewhere (Oyama, 2000) I mention the service rendered by scholars who have brought together, and so *reframed*, scattered findings on linked ecological and evolutionary changes (Gray, 1988, 1992), "epigenetic inheritance" (Jablonka and Lamb, 1995), and environmental influences on development (van der Weele, 1999). Another example is Johnston and Gottlieb's (1990) discussion of phenotypic novelty and evolution. These findings may or may not have been produced with something resembling a developmental system in mind. In each case, the relative paucity of research reflects, and helps to perpetuate, the absence of a research tradition. What work there is tends, by its very isolation, to be overlooked, and the relations among the items pass unnoticed. Strictly speaking, in fact, these workers haven't documented *bodies of research* because many scattered bits don't constitute a body. Grouping them under explicit and sometimes unfamiliar rubrics, and so suggesting novel relations among them, contributes to the constitution of a research program. Although re-framing does not automatically produce a new set of laboratory or field techniques, many innovative procedures have been devised by the people I have mentioned, and often it was precisely because they refused the usual categories of developmental influences and the usual ways of iden-tifying them.[8]

C. Where Will It All End?

Reworking categories on more than one level, opening out, and reframing all raise the issue of boundaries. Some people are troubled at the idea that an organism's skin need not be theoretically primary, but organismic iden-tity itself is hardly straightforward; consider endosymbionts, for starters. Specified limits can make the concept of a developmental system more cognitively manageable, and they can tell one where to start — or stop — an analysis. I would not, however, try to satisfy the desire for a single, de-finitive way of drawing boundaries, a catalog of the "real" constituents. My own take on line drawing reflects my generally pragmatic bent: size and shape will depend on the scope and nature of the investigation.

Many who appreciated Dawkins's (1982) book on the extended pheno-type (some of them, like me, despite disagreement with certain of his aims and conceptual means) were impressed by his skill in getting us to look at things in a different way, to divide the world up in an unaccustomed man-ner. I suggested in Chapter 8 that if one took Dawkins's vision seriously, it would no longer be possible to talk of an organism's genotype in his sense — as the genes that make it, reside in it, and make it run. "Its" genes would be spread all over the place, and their locations would be changing all the time. This point about the contours of the Dawkinsian genotype was made rather casually: Someone who has just enjoyed having her men-tal landscape tampered with is hardly going to fuss over a small matter like the stability and location of a genotype. Since then, though, the de-mand for a definition of developmental system boundaries has surfaced more than once, and I have a suspicion that the felt need for a boundary may be more acute with respect to a system of heterogeneous developmental *influences* than it has been for the outward reach of manipulative genetic *power.* I do not know how Dawkins would meet a demand to draw a line around the "extended genotype." He might not deign to answer at all, but a conceivable response from within his framework would be that it does not have a distinct outline, that it changes over time, and that it is to some extent a function of the investigator's aims. This is very much the sort of answer I give to queries about developmental systems, and Dawkins could give it for the extended phenotype itself, *if it were an issue.* Instead, it appears that the very *un*boundedness of those phenotypes is celebrated; elsewhere Dawkins (1984, p. 138) extends that potential reach to "the whole world."

Fuzzily sketched genetic entities with no fixed location, if they had to be explicitly described as such, would probably fit uneasily in the framework that spawned them, but I wonder to what extent the demand for a definitive outline is more peremptory for a developmental network than for a manipulative one. Even though the first may be described in terms of dependence while the second invites the language of agentic potency, they are themselves interdependent.[9]

Statements such as Dawkins's make a point about the way a radical reconceptualization can expand an investigative field, showing that interesting theoretical constructs don't always have the same contours as everyday objects. Any actual inquiry includes only a small portion of a huge set of relations. Sometimes research suggests a need to enlarge the portion, or an initially promising factor can turn out to be so uninformative (makes so little difference to the topic under examination) that it can be left out.

We are told that everything in the universe exerts gravitational pull on everything else. This also means everything is influenced by everything else, but this is not seen as grounds for dismay. On the contrary. When physicists engage in research and model building, of course, they opt for the manageable. In-principle unboundedness neither stalls investigation nor discounts the theory.

D. Going On

The preceding pages include several references to ways in which the conceptual resources of the developmental systems perspective are being elaborated. This final section is about how I would like to see things go on from here.

I. ARTICULATIONS

The Ontogeny of Information is not primarily a book about evolutionary theory, but it is probably fair to say that it has provoked the most theoretical engagement (as opposed to generalized approval or disapproval) from evolutionists. This is not a bad thing. One should not, however, seek to articulate the developmental systems approach with models of evolutionary processes with too great a degree of empirical specificity, any more than one should interpret it as a claim, say, about the role of *particular* environmental factors in this or that developmental process. Kim Sterelny (in press) has asked whether the approach can adjudicate several proposals about the relations between evolution and development. As a *perspective* or *approach,* though, it doesn't place the kinds of "empirical bets" Sterelny asks for. I have written on developmental constraints, one of the topics he brings up, but not in order to weigh in on one side of the debate over the relative importance of internal and external factors in evolution. Rather, I have addressed the very conceptions of development and natural selection that would cast development in the role of "inside" to selection's "outside," and have sketched a different way of relating them.[10]

The articulations I would like to see worked out are more a matter of exploring relations with other literatures, as I have done in a preliminary way not only for developmental constraints but also for structuralist biology and autopoiesis, among others (Oyama, 1999, 2000; for relations with several other projects, see Griesemer, in press; Wimsatt, 1999; Weber and Depew, 1999). These efforts help readers to see connections and contrasts among the various enterprises, but they can also deliver the occasional conceptual push to the theorists themselves, provoking clarification, revision, the sharpening or softening of distinctions.

In addition to this sorting out among research traditions, there is the articulation of basic concepts, like the notion of level. (Oyama, Griffiths, and Gray, in press, includes further examples of both, as well as further discussion of many other topics touched upon in these pages; see also Oyama, 2000.) When I referred to levels in this book I was thinking mainly in terms of size (as in molecular and organism levels) and kind of analysis (as in physiological and psychological levels). The two senses are often, but not always, associated. In more recent work I have often spoken of *scales,* to take advantage of the fact that metrics of time and magnitude are widely shared. To some extent, levels, like boundaries, are surely a function of choice of analytical category. I have not developed the tentative suggestion in Chapter 8 about ambiguous (one-many) and degenerate (many-one) mappings as markers of a shift in levels, but such relationships still seem to me to capture something significant about current usage and to provide a promising entrée into the subject. Consider the ubiquity of mapping discontinuities produced by threshold effects. The "intelligent" rock of my earlier discussion fractured abruptly under continuously increasing pressure. Recall also the striking tissue changes that can result from rather simple cell movement.

Though we can often do without formal definitions, a sharper sense of levels could clarify relations *among* treatments. As we saw earlier, one can compare theorists with respect to the level to which they drive a parity argument. To take another example, much of Elman et al.'s (1996) connectionist "rethinking" of innateness in cognitive psychology depends on distinguishing levels of cognitive processing. In today's heavily nativist cognitive science and linguistics, rethinking innateness is no mean task. These authors make some acute observations about the importance of developmental change in cognition but insist that they are not antinativists (p. 46). Indeed, while they restrict notions of genetic prespecification and preformation more than most of their colleagues, they are squarely within the conventional "interactionist" tradition described earlier. They dispute certain catalogs of innate features but to a significant extent leave the innate-acquired distinction alone. Most of their arguments about connectionist models do not require full embrace of the constructivist interactionism they cite with approval, just as Sidanius et al. (1991) did not need to embrace it in order to make some of their points. These scholars' critiques of a variety of nativisms, however, are not as telling as they might have been if they had taken their interactionism more seriously—if they had applied it at more levels, if you will. My discussion in Chapter 7 of Raff and Kaufman (1983) describes a similar situation.

Other pivotal concepts could be clarified through (constructive!) interaction with other fields. Relations between the organism-environment complexes of the developmental system and ecologists' understandings of the niche are a case in point.[11] Recent philosophical comments on developmental systems work, furthermore, suggest that constructivist interactionism brings up quite general questions about the meanings of causal "determination" and explanation in the life sciences.[12] Additional clarification on this last front might reduce the frequency of the sorts of misunderstandings that have been mentioned here. I hope so. Seeing one's work cited for the wrong reasons is a distinctly mixed pleasure.

2. RETICULATIONS: MOLECULES, MINDS, AND COMPUTERS

Ideas of representational correspondence loom large in this book—correspondence between genetic "information" and the processes or outcomes of phenotypic development, or between a phenotype and its surround. Sometimes more conventional notions of mental representation appear, as in Lorenz (1965).[13] When we look at the gene, we look inward, toward the mystery hidden inside every cell. When we "look" at the mind, we look inward in a different sense; and when we look at computers we don't think of it as looking inward at all, but in a sense we are. The linked histories of molecular biology, cognitive science, and computer science trace our ways of dealing with complicated processes whose inscrutable creativity seems disproportionate to their relatively modest means: bits of silicone, neural tissue, macromolecules. Encoded representations promise an explanation. Earlier we looked at Dennett's and Chalmers's treatments of genes and mind and saw that both authors draw on the densely reticulated set of images and meanings examined in the preceding chapters. We *make* computers in a way we don't make genes or minds (this is changing for genes) so perhaps it is fitting that "programming" comes to serve for all three. As I suggested in this book, we draw on our experiences as makers in trying to deal with such cases, though metaphor's arrow points both ways. In the standard view, DNA molecules "process" whatever "information" they need to in order to make us and to make us run. By a neat doubling back, we are now sometimes said to "program" our children; cults, religions, and sometimes whole cultures are supposed to do the same for/to their members. It's worth a moment's thought that all of these programs connote some degree of mindless compliance. This may account for some of our difficulty in working them into satisfying accounts of experience and agency.

Molecular biology is a significant anchor for this complex, so it is interesting that contributions to the critical reenvisioning of the gene seem
increasingly to be coming from there.[14] It is toward the study of mind,
however, that I am currently looking with special anticipation. This book,
which is in some sense about how we think, generally treats the mind
itself as given. Yet I have long suspected that the perspective it presents
is relevant to mental processes, which are, after all, part of an organism's
relations with its surroundings, and with itself. I have also maintained
that the apparently transparent notion of an organism's being in-formed
by preexisting "information" is as problematic for the environment as it
is for genes. Hence my interest in the explanatory burdens placed on information by people like Dennett and Chalmers.

Today's cognitive science is heavily "cognitivist" — the mind is treated
as a machine for manipulating symbols, an information processor firmly
ensconced in, or identified with, the individual brain. Mentality is increasingly being cast in different terms, however, of sociality, activity, and
physicality; of distributedness, embeddedness, and embodiment.[15] These
diverse unorthodoxies may well contain means for constructing a different approach to life and mind, one that includes all the stuff, the activity,
and the relations for which "information" must substitute once they are
stripped away in the formalizing process. Much of my unease with mainstream cognitivism, in fact, is of a piece with the preferences and skep-
ticisms that color my approach to much of mainstream biology. In both
we encounter a tendency to interpose a screen of representations, information, or instructions between interactants and interactions, between
organisms and their environments (including bacteria and their wet rags).

There are some parallels between certain aspects of my thinking about
these matters and John Searle's (1992) critiques of cognitivism and the
mind-matter duality.[16] If I had a full-fledged view of cognition, I doubt
it would resemble his, but whatever his other commitments, Searle does
make some barbed queries (p. 199) about the insertion of a "level" of
symbol manipulation between brain processes and knowledge that seem
very much to the point.

In Chapter 7, I denied offering "some astounding new fact that requires
discarding large chunks of theory," and listed some apparently modest
goals. I meant it, but I had also been pursuing those goals long enough

to know that once I was past the "Of course" stage with an audience, the perceived scope of my argument was apt to change. I don't know when the pursuit of consistency and clarity becomes something else. I am still impressed by the frequency with which the ideas in this book are described both as worn-out truisms and as too wild and crazy to contemplate. Perhaps the continuing inquiries I've pointed toward in this concluding section will change this. Perhaps not. In any case, I think there are interesting times ahead.

Notes

1. Introduction

1   In her discussion of theories of generation from the late seventeenth to the early nineteenth century Gasking points out that a preformationist attitude does not require outright encapsulation (1967, p. 91). See also note 2, chapter 3.

2   Again, vitalism is not a unitary, simple, or clear notion. Workers like Buffon and Maupertuis attempted to relate their internal molds and forces to physical forces, regarding them as causal and explanatory. The problem was to show how organic form arose, and notions of specific forces acting on identical particles, or of universal forces acting on differentiated particles, were attempts to address it. These scientists reserved the charges of vitalism and occultism for their opponents while defending their own views as scientific (Gasking, 1967, pp. 83–96). Jacob (1973, pp. 74–82), argues that Newtonian mechanics required a notion of hidden structure, of organization as the key to understanding three-dimensional complexity, and that the popularity of elementary units in late eighteenth-century biology reflected a preoccupation with finding the biological equivalent of atoms and their attractions. A nonpreformationist explanation, then, required a "hidden order behind the visible order," consisting of arrangements of units. This idea derived directly from the corpuscular view of matter and was as much the result of logic as of empirical considerations. The organizing forces, then, whether actually conceived of as "mental," as in Maupertuis' theory, or as more directly analogous to atomic attraction, as in Buffon, had to serve the function of an organic memory, allowing the assembly of particles into living organs and beings that would perpetuate the forms of their parents.

3   Early preformationists, in fact, ultimately relied on divine power to create all the nested forms, so the dispute in some ways was not over whether form was created, but when, how, and how often it was created. Some developmentalists use "epigenesis" the way I use "ontogenesis" here (Oppenheim, 1980; Waddington, 1975) while others eschew the term because of its association

with environmental determinism (Schneirla, 1966), and still others co-opt it for a variation of biological determinism, albeit a much elaborated one (Lumsden and Wilson, 1981). To complicate matters further, Gruber associates "ontogeny" with the species-typical developmental pathway (1980), whereas I have suggested that "maturation" be cleansed of its genetic determinist and physiological overtones and applied to such typical pathways instead (1982); this would free "ontogeny" to refer to development in the most general, inclusive sense, the sense in which it is used throughout the present book. For some historical discussion of these issues, see Gasking (1967), Gould (1977), Jacob (1973), and Oppenheim (1982).

4 In *Not in Our Genes,* Lewontin, Rose, and Kamin (1984, p. 268) call "interactionism" the "beginning of wisdom," but go on to reject it for sharing many of the defects of genetic and environmental determinism. A strategic decision must always be made with respect to such ill-defined but influential complexes of ideas. One can avoid inflicting another *ism* on one's public and try to rehabilitate the term in use with criticism and reformulation, or one can abandon it and attempt to introduce a new one instead. The former carries the risk of seeing one's alterations ignored or reinterpreted to fit the existing constructs. (This has happened often enough with interactionism that one could argue that I should have been deterred from choosing this strategy by the fruits of my own analyses.) The latter choice proliferates labels that are sometimes difficult to distinguish, and that, again, may be ignored or misinterpreted. Though I have taken the first route while Lewontin et al. have chosen the second, there is much more that unites our critiques than divides them, and the success of our shared endeavor means more to me at this point than our possible disagreements or the labels that ultimately prevail.

Much the same can be said about the concept of integrative levels, described by Tobach and Greenberg (1984) in a volume dedicated to T. C. Schneirla. I agree both with their criticisms of much "interactionist" thought and with their prescriptions for correction, but have not yet given up on the term. I shall distinguish the sense of interactionism I favor by using a modifier like "constructive" or "constructivist," and by using scare quotes for the more conventional usages.

2. The Origin and Transmission of Form:
The Gene as the Vehicle of Constancy

1 Because the ideas I criticize in this book are dominant ones in a number of fields, I will often be disagreeing with people of great achievement, whose

work I admire and depend upon. I trust that such disagreement will not be mistaken for lack of appreciation.

2   Again, the structure of the arguments is familiar. In the early 1900s, Netting (1977, p. 3) reports, the anthropologist Alfred Kroeber described the role of environments only as sources of limiting conditions for human societies, not as "active or positive causes." He presumably reserved the latter function for culture. Similarly, behavior theorists, when they have acknowledged the importance of species differences or motivational effects, have tended to treat them as necessary conditions and constraints: that is, as sources of raw materials and boundaries. Since Hinde and Stevenson-Hinde's (1973) *Constraints on Learning,* in fact, "constraints" has often served as shorthand for "phylogenetic differences." It is, of course, in the construction of morphology, physiology, and instinct that genetic causes are seen as acting within environmental constraints and on nongenetic raw materials, while in learning it is the environment that imposes form, assuming only a normal nervous system and thus a full set of reflexes and operants. In neither case do careful writers deny the necessity of the factors described as supportive, permissive, or constraining; nor do they deny that these factors may contribute to variance. A distinction is nevertheless made between control of essential form and trivial detail, between true causes and mere constraints.

3   Toulmin (1970) attempts to think about causes and reasons without either opposing them or assimilating the latter to the former. Though I suspect that his reliance on a distinction between learned and unlearned in defining reasons is at best a provisional one, he makes an important point when he says that claims about reasons and causes are quite different kinds of statements, made on different levels. Similar attention to the concept of blame seems warranted. It might be argued that there are other ways of dealing with blame than by casting oneself in the role of a will-less object.

4   Hyland (1984) discusses the confusion over the internal-external dimension in personality theory. His critique of "interactionism" makes it clear that the conceptual and methodological parallels between the persons-situations imbroglio and the nature-nurture debate are even more numerous and striking than my brief treatment can describe. I recently heard a geneticist describe the distinctive, "autonomous" properties of a cell as the cell's "personality."

3. The Problem of Change

1   Because ontogeny continues at least to the adult period—and, some would argue, through senescence to death (Bonner, 1974, pp. 167-168, for instance)—what is considered the "end" is somewhat arbitrary. Though the

adult is traditionally seen to be the end of development, both in the sense of goal and in the sense of terminus, with everything else either an incomplete transition to, or a degeneration from, this basic form, a strong argument can be made that ontogenesis is continuous with the life cycle. Every stage is thus equally the "end" of development. Unlike a machine, which is generally useless until it is completely assembled, an organism "works" at all points in its development. For a view of development that places ends in genomes, this implies that not only the standard adult form but also every stage of life cycle must be "in" the genes. When norms of reaction or sets of potentials are thought of as being encoded, the genomic freight rapidly becomes prodigious.

2   Gasking distinguishes between preformationism as a *prediction* of what would be seen in the germ, especially if parts were hardened and made visible, and as an *explanation* of development. It was the latter that was generally primary, and it was quite consistent with gradual appearance during embryogeny (1967, pp. 48–50).

3   These points become especially interesting and important in Gould's case, for unlike many of the other writers cited in these pages, he has been persistent in his opposition to various kinds of genetic determinism and alert to their consequences (1981, and numerous columns in the magazine *Natural History,* for instance). Over the last several years his writings on these issues have become steadily more "interactionist." Program metaphors seem to have dropped out, and in their place is an opposition between biological determinism as a "theory of limits" and "biological potentiality" "viewed as a range of capacity" (1984, p. 7). For reasons that will become clear later, this formulation unfortunately does not quite solve the problem. It neither distinguishes his position from those of the workers he criticizes (many of whom use an "interactionist" vocabulary of sorts and readily speak of *potential* rather than *fate*) nor, much the same thing, fully detaches him from the notion of fixed limits (often expressed, in fact, as biologically encoded potential). What are limits, in fact, but the boundaries of the range of capacity? (See Chapter 5, this volume.) Gould's eloquence and moral commitment are considerable; we will all benefit from his continued thinking about these matters.

4   In *The Ghost in the Machine,* Arthur Koestler gives a hierarchical systems account of ontogeny, phylogeny, and creativity, covering some of the problems treated here but in the service of a quite different aim. Having been accused of criticizing behaviorism once too often, he advocates the formation of the spcdh, the Society for Prevention of Cruelty to Dead Horses (1967, p. 349). For an interesting treatment of Koestler's work, see the three reviews by Stephen Toulmin collected in Toulmin (1982).

5 It is perhaps significant that the arguments the preformationists had used against each other in the ovist-animalculist controversy were deployed in the mid-eighteenth century by Maupertuis against the preformationists themselves (Gasking, 1967, pp. 70–78).

6 "Active" and "passive," as should be clear from this discussion, do not generally characterize entities, but only signal focus on sources of perturbation and effects of perturbation, respectively.

4. Variability and Ontogenetic Differentiation

1 The model of variation from a constant genotype in (1) and (2) is oversimplified, since gene replication is not always unerring. Somatic mutations seem to be crucial in the immune system. Still, I am not aware of serious challenges to the generalization that differentiation involves differential gene transcriptions rather than systematic alterations in the genetic material itself (but see Van Valen, 1983, for a dissenting view). Changes in the gene pool by a number of mechanisms are, on the other hand, basic to evolutionary processes. The genotype "drawn" from the pool by an individual is in some ways conceptually similar to the subset of genes that is biologically active in any given cell of an organism at any given moment.

The structure of the three kinds of radiation may be more similar than it seems. While ontogenetic radiation or differentiation occurs within the same skin and is therefore integrated in its physiological and mechanical interactions, related taxa may interact ecologically, and naturally occurring clones may interact as well, influencing each other's development. The boundaries among individuals, and therefore between ontogenetic and phenotypic radiation, are further obscured in colonial organisms of various types, especially those that aggregate only under certain conditions or during one portion of the life cycle. (See also P. P. G. Bateson's 1982 implication that an individual might even include different species in close contact; Gould, 1982; Hull, 1984.)

2 According to Waddington, mice raised in warm temperatures have longer tails than those developing in cooler temperatures. The extra length may enable them to dissipate heat more rapidly (1960, pp. 132–133). Mice are found in many habitats, but if they tended to live only in hot climates the long tails would be species-typical and inherited in one traditional sense. Shorter tails in the individuals who were raised in the cold would be abnormal, "exogenous adaptations," "phenotypic adaptations," acquired traits, and the like.

Certain insects feed on toxic plants and become toxic themselves. Their toxicity is species-typical because their feeding habits are (but see M. D. Bowers, 1980), and tends to be advertised by conspicuous coloration (Brower and Glazier, 1975). The role of such chemicals in the lives of these creatures does not in any way require that the toxins be conventional metabolic products; the predator is indifferent to such niceties, and so, surely, is the selfish gene.

5. Variations on a Theme:
Cognitive Metaphors and the Homunculoid Gene

1 The idea that knowledge is a commodity of which minds can have more or less makes it difficult to think reasonably about mental processes. I suspect that we are better off thinking of knowledge as something minds *do,* rather than as something they *have.* In the same way, it is often more useful to see information not as a set quantity that can be stored or transferred, but as a kind of description of events and processes as seen from a certain point of view.

2 It seems odd that epigenesis and canalization, notions used by Waddington (1957, 1962, 1975, for example) to elaborate an extraordinarily subtle interactive view of development, in which there was no room for discrete ontogenetic sources of any sort, and in which the misunderstanding of genetic control and inheritance was repeatedly warned against, have been adopted by authors who conceptualize environment and genes as competitors in a zero-sum game for control of "knowledge" and of the "behavioral repertory" (Lumsden and Wilson, 1981, p. 332). This conception, and much else that is found in *Genes, Mind, and Culture,* is alien to Waddington's style of thought.

3 We again see the confusion between *change* within an individual (which requires a succession of transitions from state to state, each state based on environmental as well as genetic "information") and *variability* among individuals (which involves not more or less input from either source, but different input).

4 Mayr says that the progression of an ecosystem is not attributable to a program, but rather is a "by-product of the genetic properties of the constituent species" (personal communication, April 1983).

Hofstadter, drawing on E. O. Wilson's work, describes the complex functioning of an ant nest, pointing out that an individual ant's small brain can't be thought of as containing information on nest structure. The "information" must be dispersed—through the age and caste structure of the colony, as well as through their bodies and elsewhere. "That is, the interaction between ants is determined just as much by their six-leggedness and their size and so on, as by the information stored in their brain" (Hofstadter, 1980, p. 359). He devotes a chapter (chapter 10) to discussing levels of analysis, movement, and confusion among levels, and the problems of dealing with multileveled nested structures. The operating characteristics of each level may be very different, may be intimately tied to their physical structure, and may be impossible to express at a lower level, a fact Mayr acknowledges in a way when he describes a reproductive pair as a unit.

Because "programs" for such complex systems do not reside in the genes of their constituents, any more than programs for ocean currents reside in molecules of water, but are rather our attempts to formalize complex systems (and are thus in some sense descriptions of our own activities and knowledge as well), there is no barrier to treating as a unit of analysis a system that includes several individuals of the same species, individuals of different species, or organisms and the nonliving environment.

Gunther Stent has described ecological progressions as the consequence not of "an ecological program encoded in the genome of the participating taxa" but of "a historical cascade of complex stochastic interactions between various biota" (quoted in Lewin, 1984, p. 1328). Again, Lewontin (1982, p. 161) describes the ways organisms both create and destroy their own environments. When he speaks of ecological succession as "a consequence of the alteration of soil texture, chemistry, moisture, and light created by assemblages of plant species to their own detriment" (certain trees create such dense shade that their own seedlings cannot survive, for example, or consume resources at too rapid a rate and are replaced by another species), he could just as well be referring to some of the ways cells and tissues alter the conditions for their own growth, proliferating, moving, changing shape and function, or dying off as a result of changes they have helped bring about. In speaking about programs, then, it would appear that the useful question is not, Where is the program that makes this system work? but rather, What is the system I wish to describe with my program? Then, even if one's program is identified with the operating characteristics of the system being described, one is not leaping across so many levels that intelligibility is lost, and with it the possibility of describing anything at all.

Lewin gives an excellent account of the decidedly nonreductionist lessons to be drawn from even the most reductionistic analyses in developmental biology, including those of a peculiar worm that always has the same number of cells in each of its organs and in its body as a whole. (Even the most apparently clockworkish developmental systems on closer inspection seem more like Rube Goldberg devices; the article is entitled, "Why Is Development So Illogical?") In the course of his exhaustive studies of the derivation of each of the 959 cells of the nematode worm, *Caenorhabditis elegans,* Sydney Brenner has given up his goal of discovering the genetic program that would explain this miracle of developmental precision and now warns against even metaphorical use of program terminology. Because evolution is changes in DNA, however, Brenner nevertheless believes that the "total explanation of all organisms resides within them," though he acknowledges the difficulties this presents for understanding development (Lewin, 1984, p. 1327). Because he does not question this assumption about the relation between evolution and development, he must continue to look for an "internal representation" or "description" inside the organism. This is precisely the shoal on which attempts to relate ontogeny and phylogeny have foundered again and again. See discussion of Raff and Kaufman in Chapter 7.

5 Notice this is considered an open program only because of the peculiar manipulations that experimental embryologists have performed in their attempts to understand ontogeny. In the absence of environmental transplanting agents, this induction is extremely reliable, involving little or no variability due to the "insertion" of environmental information, and is thus a prime candidate for explanation by closed program.

6 For a sophisticated version of this argument, see Fodor (1980). He seems to conclude that the only way out of the nativist impasse is to reformulate learning, which he treats as hypothesis formation and confirmation. I would suggest that it is not only learning that must be reformulated, but all of development, since, as we can see, traditional conceptions of ontogeny lead to the same paradox as do traditional conceptions of learning.

In a discussion of brain states, Hofstadter points out that a given state has alternative pathways, but that information specifying which one will be taken is not in the brain, but depends on external circumstances. Both brain states and circumstances must be "chunked" (summarized) probabilistically, and with reference to each other. He notes that specifying a person's potential thoughts or beliefs without reference to possible contexts is as senseless as trying to refer to the range of his "potential progeny" without considering his mate. One cannot speak, then, of all possible thoughts being dormant in the brain; how could one imagine a decision procedure for distinguishing possible thoughts from impossible ones (1980, pp. 382–384)? The parallel to norms of reaction and information on possible developmental interactions, contexts, and outcomes is clear.

7 Both Monod (1971, pp. 60–61) and Jacob (1973, p. 279) compare enzymes to entropy-reducing demons, and G. C. Williams refers to the Darwinian demon that maximizes adaptation (1974, p. 89).

8 I don't think it accidental that such ambiguity exists about the receiver of

information. Specifying it would force us to deal with the relativity of the amount and meaning of information to the "decoding mechanism." This in turn would bring up all the problems discussed in this section. The information in a stretch of DNA, for example, has different meaning for the phenotype at different stages of development and under different conditions, and it has different meaning in different tissues at any given point in development. The same may be said of environmental information. The consequences of an event, and its meaning, depend on what is going on, as well as how much and what you know. This varies with level of analysis and developmental state and is thus relative to the question being asked.

9 It is interesting, incidentally, to compare Schrödinger's images of architects and builders (1967, p. 23) with sixteenth-century descriptions of the generation of living beings by the uniting of form with matter (Jacob, 1973, pp. 23–25). These descriptions are full of images of artists and builders, including architects and carpenters; Jacob observes, "heredity represented the share of the artist, as it were, that mixture of form, constitution and temperament but not matter, which through seed reappeared from one generation to another."

10 Dretske (1983, chapter 1) discusses the difference between information and causation.

11 Among social scientists there is occasionally some concern with distinguishing cultural from biological universals, a matter to which we will return. This concern assumes the possibility of allocating causal efficacy among factors that do not vary.

6. The Ghosts in the Ghost-in-the-Machine Machine

1 I think the problems Ryle found with this kind of explanation are quite real, though I have reservations about some of his solutions. See Sober (1983) for some comments on mentalism.

2 See papers in Brim and Kagan (1980) on the continuity-discontinuity problem in developmental studies. I was struck by the parallels between parts of these discussions and my own. This makes sense not only because we are engaged in similar sorts of inquiry into rather large, sprawling, historically complex issues, but because the issues themselves are intimately related. One of the most troublesome aspects of the nature-nurture debate has been the conflation of "genetic influences" with some sort of developmental continuity and of "learning" or "environmental influences" with discontinuity.

3 At times the genes even seem to be endowed with emotions and motives —

competitiveness, selfishness—and their relevance to our own existence is often relegated to the field of the emotional and irrational, the unruly animal core with which each of us must construct an uneasy truce.

4 When we add to this package of perplexities the fact that the gene is customarily given the role of the unmoved mover (Campbell, 1982, afterword), which puts it and its products largely beyond our control or influence from the outside (while what is environmental, psychological, or social is assumed to be manipulable at will), the need for clarification becomes pressing indeed.

5 "Emergence" is a loaded word in many circles, implying quasi-mystical appearance of something from nothing, or at least, from not enough. Perhaps we need another term to express the progressive manifestation of form by the more or less orderly operation of interactants, all of which may influence the developing product, but whose precise significance in a particular interaction depends on the characteristics of that interaction. "Ontogeny" would seem a candidate term, but it will serve only if detached from all manner of ghostly engineers and the presumption of uniform preset developmental courses (note 3, Chapter 1).

6 Even the ovum undergoing fertilization, long an archetype of biological and psychological, especially female, passivity, is seen to be extraordinarily selective and active in arranging and effecting its fateful meeting (Schatten and Schatten, 1983). This description is also a useful antidote for the image of the egg as a bag of chemicals.

7 For the same reasons that it is folly to partition mental characteristics, it is folly to partition morphological or physiological ones. Yet many workers manage to apply this interactionist reasoning to behavior, especially human behavior, while continuing to speak of "physical" traits as genetically determined or programmed. This suggests that they have not applied the *reasoning* at all but are finding another way to say that behavior is controlled by the genes less than the body is.

8 E. O. Wilson cites the same research and draws the same conclusions (1978, p. 100). He switches repeatedly, in this book and elsewhere, among definitions of innateness and genetic determination by heritable differences within populations, by genetic differences among populations or species, and by probability that a trait will develop in certain environments (chapters 2, 5, and glossary, for instance).

9 Here he refers to Waddington's genetic assimilation work on wing veins in flies, but where Waddington always rejected the dichotomy as it is generally understood and tended to use quotation marks around "inherited" and "acquired" (1957, p. 168; 1975), Konner uses none.

10 Church law is not ordinarily thought of as hypothetical, and therefore open

to revision on the basis of data, though it may be *Wilson's* belief that moral law should be so construed.

11 He seems to disapprove of violence and unrestricted reproduction in an overpopulated world, for instance, and presumably has no qualms about disliking "cheats, turncoats, traitors." He considers the latter dislike, in fact, to be based on "innate learning rules" (1978, p. 162), even though, in some sociobiological theory, "cheating" can be under selectional pressure, just as "altruism" can.

12 Things are even more complex than they appear in this presentation. While what is not theologically natural may be "unnatural," and therefore undesirable in such a system, what is not natural in Wilson's sense encompasses not only the pathological but all "cultural" or "nongenetic" phenomena as well. Even if these latter are conceptualized as mere variants on biological themes, they are not on this account worthy of disapproval or approval. So the problem of defining the undesirable remains. Is it possible, adaptive utility aside, that Wilson believes that because homosexuality may be genetic, it is beyond individual responsibility, and therefore not subject to moral disapproval? But if hatred of cheats is based on innate learning rules, which are adaptive, might not hatred of sexual variants be similarly "biologically" sanctioned? How would one tell, and what implications would this have for our morals?

13 See especially her chapters 10 and 11 on the significance of human language and on the "colonial picture" of reason as governor of chaotic passions.

14 Konner does not say in this passage that he believes that this "biological" gender difference exists, but an anecdote in the preface makes it clear that he does.

15 The precise meaning of this term is not agreed upon by scientists (Lerner, 1980). Fishbein says the "interactionist hypothesis" is that psychological development is a function of maturational level and environmental influences in interactive, not additive relation (pp. 58–59). If "interaction" means only that an environmental influence may have different effects at different times in development, the most this view offers is an emphasis on a generally accepted, though important, idea. The notion of maturation, in fact, is partly defined by the changing nature of environmental relations (don't try to teach calculus to a three-month-old). One would have thought that one of the important insights of interactionism was rather that the mutual dependence of organism and environment disallows both the traditional concept of maturation and the language of designer genes.

16 See Waddington's disagreement with conceptions of "non-genetic plasticity of the phenotype" (1975, p. 89) and his definition of "inherited" and "acquired" strictly in terms of frequency of appearance of characters in particu-

228 Notes to pages 116–121

lar sequences of environments. Unless one follows his definitions carefully, it is possible to be misled by his rare ambiguous references, like "acquired characters being converted into inherited characters" (p. 61) or a phenotypic character "first produced only in response to some environmental influence" later being "taken over by the genotype." Even this last statement is immediately explained: "so that it is formed even in the absence of the environmental influence which had at first been necessary" (p. 91). To Waddington, every phenotypic character is a joint function of genotype and environment. He uses "genetic control" only to indicate variation with genotype under specific conditions.

Though frequently cited by those who are attempting to synthesize the biological and behavioral sciences, Waddington's work, especially that on genetic assimilation, is too often taken as an opportunity to retain the inherited-acquired distinction while establishing interactionist credentials. The point made in the penultimate sentence of the previous paragraph was already a truism when Waddington made it in the 1950s and 1960s, but the *point* of the point has still not been made, even to many of those who use his work.

17 It is worth noting here that Piaget, one of the thinkers Fishbein says most influenced his approach, would hardly have considered hunter-gatherer cognition as more basic or representative of human biological endowment than that observed in Geneva; his interest was always in universals.

18 This example is chosen less for its currency than for its utility in illustrating the combination of analytic good faith and faith in innate form that is the subject of this discussion. I would argue, though, that its differences from many contemporary efforts are fairly superficial, while its similarities are deep and important. The unlearned as biological substrate persists in the literature.

19 Harris resists reduction of most cultural constancy and variability to the biological. He denies, with minor qualifications, genetic differences as determinants of cultural differences, which he does not distinguish from genetic programming of characters. Instead of using an interactionist argument to combat genetic determinism, he accepts genetic determinism for some things and argues against it for others. He attacks some forms of interactionism, in fact, and interprets the objection to partitioning characters as one partitions variance as a claim that "cultural and biological influences equally determine cross cultural diversity" (1983, p. 21). This kind of confusion does not help his case against misuse of biology.

The fact that so many critics of sociobiology share its basic assumptions about the nature-nurture distinction and about dual developmental processes weakens their criticisms, either reducing them to quarrels over how much and which behavior is biological, or rendering them seriously incomplete. This is the case with Sahlins's otherwise interesting critique of sociobiology (1976).

20 Precocity and stability are themselves not necessarily related, though this is a hotly contested and complex issue among developmentalists; operationalizations of characteristics and of stability become crucial (Brim and Kagan, 1980; Hinde and Bateson, 1984).

21 In a later paper Plomin (1981) shows considerably greater circumspection in using heritability, which he notes has gained a bad reputation. He nevertheless regards the "mistaken interactionist view that the separate effects of heredity and environment cannot be analyzed" as senseless as the naturenurture issue itself, since both are needed for behavior (pp. 258, 260). He defends behavior genetic research and cites a finding from the Plomin and Rowe research as an example of an interesting result. In a group of young twins, social behavior with a stranger showed some heritable variation in some test situations, while behavior with the mother did not. As an interactionist of sorts whose discomfort with the nature-nurture issue goes far beyond Plomin's but who has never argued that behavior genetic analysis is impossible or necessarily pointless, I also find this an intriguing result and would like to know more. But I wonder whether behavior toward strangers in certain situations is now to be considered part of the core of human personality while behavior with the mother is not, and just what this means: Will it show more long-range continuity? Will it order other aspects of interaction in a more significant way? To what extent is it peculiar to the psychological dynamics of twins? and so on. I wonder, too, whether this finding will lead to closer study of these aspects of early sociality or whether, as has too often been the case, the quantification of heritability was the real goal.

That a treatment of temperament is possible without such definitional contortions is evident from the work of Thomas and Chess (1977). The refusal of these researchers to claim genetic status for their constructs does not reduce the usefulness of their findings in understanding ourselves as biological beings; in fact, they are that much more valuable in illuminating the manner in which real living beings differentially shape and are shaped by their surroundings. See also papers in Lerner and Busch-Rossnagel (1981).

22 We are left with a biology that defines the boundaries within which free will may operate, if it may operate at all, and that conceives of a responsibility that exists only in the residual space that remains after causal explanations have done their best, or worst. It allows us to ask questions like, Do humans have free will (do they act for reasons?), or are they determined (do they act by causes?)? a logic that also tends to oppose humans with minds to mind-

less animals or to require that if humans are "genetically programmed," then they are nothing but animals or machines. This in turn encourages a second-level opposition between the rational and free in man and the beastly biologically driven in him. The unity of the developmental system is thus compromised by dual ontogenetic processes, one to produce Dr. Jekyll and one for Mr. Hyde. In such a view of development, epigenesis and preformationism are not synthesized, but are used in tandem to satisfy whatever version of developmental duality is in vogue.

Current treatments of animal cognition are rife with oppositions of thoughts to genetic programs, plans to instincts, anticipation to hard-wired instructions (Griffin, 1982, 1984). Though Griffin and others are to be commended for reopening the long-taboo topic of thinking in nonhuman creatures, and though Griffin himself questions the utility of assuming that genetic programming and mind are incompatible (1984, chapter 5), much more must be done to break the Cartesian deadlock over minds and machines that threatens this area with stultification before it has gotten its bearings.

Innovative, subtle treatments of difficult issues often don't get very far because colleagues and the public cast statements and findings in matter-mind, nature-nurture terms. Witness the fate of a Schneirla, who is sometimes described as a behaviorist or an environmental determinist, or a Waddington, who becomes an unwitting ally of the very genetic determinism he criticized. Beer (1975b) has written a fascinating and quite moving account, "Was Professor Lehrman an Ethologist?," of some of the personal, ideological, and terminological issues at stake in the relations between traditional European ethology and American students of animal behavior. In it he touches on some of the difficulties of moving past these dichotomies even when the goal of synthesis is shared. It would seem that if we want to make progress in thinking about thinking, we would be well advised to devote some thought to what "programming," "machine," and their supposed antitheses mean (Beer, 1983b).

The futility of trying to decide whether to attribute complex behavioral coordination to minds or genes, of course, is revealed when we see that attributing it to the genes ultimately reduces either to postulating a mind in the genome or "simply" to describing the system that produces the behavior. It is the latter that is the real point of the inquiry, and if it is accomplished then there is no need to choose between minds and genes or to figure out how learned actions are combined with innate ones.

23 Chomsky (1972) is steadfast in his commitment both to the innate and to radical politics. Many other notable exceptions to the nativist-conservative, empiricist-liberal generalization exist, and the ways all these people connect their science to their politics is an interesting question in itself. Liberal

colleagues have told me that their political views *require* them to reject biological approaches, and one suggested that it was better not to know the biological truth that might threaten his ideology. I submit that it would be better to question the traditional meanings of the biological to which both liberals and conservatives respond. One thing that distinguishes Chomsky from many other nativists is his concentration on the rational and uniquely human rather than the irrational and phylogenetically common. Notice that this approach avoids the problem encountered by many sociobiologists, namely, that they are often charged with legitimizing the beastly in humans with their science, and thus must counter the charges by urging us to transcend our own nature. Given the view of that nature as ruthless, calculating, and selfish, and given the belief that deviations from nature are possible only at great risk and effort, the veneer of human virtue in these accounts is won at the price of constant vigilance and considerable stress. A possible consequence of this view of life is that, by fixing our attention on the ever-present evil in the shadows, we become enthralled by it. It defines what we become whenever we are weak, discouraged, angry, defiant, or simply on our own. The self-fulfilling prophecy is devious in its workings.

Harris (1983) claims that the separation of cultural from genetic programming of social life is imperative because our democracy is based on the importance of differential socialization; sociobiology is therefore open to charges of racism. Peterson and Somit (1978), on the other hand, observe with approval the incorporation of biological concepts into political science, suggesting that sociobiology may resolve the twenty-five-century-old dispute over the nature of political man. They say that "the genetic basis for altruism in humans suggests an inherent goodness" (p. 454), then quickly allow that competitiveness and aggressiveness seem also to be natural. They note that sociobiology appears to support a conservative view that behavior is rooted in the nonrational and that individual differences in ability require superordination and subordination in politics, then immediately add that reciprocal altruism supports a liberal, free-market economy. Far from raising doubts in Peterson and Somit's minds about the relevance of a field that can give such undiscriminating support to opposing views, this interpretational freedom seems to be seen as grounds for enthusiasm. In an academic world in which the prestige and credibility of a science seem to vary inversely with its distance from physics, biology is most scientific when it is most biochemical. The grounding of evolutionary studies in the molecular biology of the gene, by means we have examined in these chapters, gives social scientists a new and powerful way of claiming legitimacy: explanation by evolutionary adaptation.

24 Larry Niven and Jerry Pournelle wrote a book called *The Mote in God's Eye*

(1975). It describes, among many other things, a kind of being, affectionately referred to by humans as "brownies," that in many ways exemplifies the characteristics that are attributed to the genes. Brownies are very small, nonintelligent in a conventional sense, but incredibly dextrous, with a purely instinctive mechanical ability that allows them to analyze, disassemble, and build or rebuild anything in a completely organic, integrated, efficient, and multifunctional design. They reproduce at an explosive rate, organizing the environment to their own advantage while seeming to help humans in many ways. (They silently repair defective objects at night; hence, "brownies.")

In one scene, humans in space suits leave a spacecraft in an attempt to escape the brownies, which they now see not as cunning pets but as a mortal threat. One man sees a space suit approaching and assumes that a comrade is inside. A glance through the faceplate, however, reveals a number of minuscule brownies, manipulating the controls and limbs of the suit, mimicking intelligent human behavior, and propelling the whole cumbersome affair through space. This image, which comes to my mind whenever I think about the homunculoid gene, should be compared to Richard Dawkins's amusing description of genes driving a huge, clumsy robot across the landscape (1976, p. 21).

Starting with the gene as the unit of selection, and therefore the unit of selfishness, commits Dawkins to seeing all adapted gene effects as selfish, by definition, no matter what strategy is used. When some aspect of an organism's behavior or physiological functioning seems not to be in its own genes' interests, it thus suggests manipulation by another creature's genes. The idea that there can be *units* of selfishness, and that these can manipulate organisms and the inanimate world, obliges Dawkins to view most phenotypic features as evidence of either gangsterlike ruthlessness or impotent suckerhood (manipulation by an organism's own genes or by another organism's genes).

7. The Ontogeny of Information

1   This does not keep some commentators on cybernetics and computer science from speaking of programming as unidirectional control. Foster (1975, pp. 123–124), for example, describes two lines of communication between a computer and the system it controls, one for sensing and one for activating. Both lines, he asserts, "belong to the computer," and both, whether used for inquiry or for control, are therefore "fundamentally *outgoing*." This produces a master-slave relationship because the process being controlled is not "aware" of the computer. One might argue, actually, that one of

the points of the master-slave relationship is the latter's constant awareness of the former; such relationships often carry the possibility of subversive counterinfluence.

2 Having contrasted such inherently precise mechanisms with conventional machines and pointed out the ways in which the former are more fitting models for many biological processes, Elsdale does say that since the rules constraining cell interaction are under "direct genetic control" (1972, p. 101), they don't vary. Yet presumably cells of other tissues can contain the same genes without behaving the way these fibroblasts do, and even fibroblast behavior varies with conditions (and presumably with developmental state), as Elsdale himself shows. Thus, if by "rules" he means "mode of activity," the rules do vary. If, on the other hand, rules are higher-order contingency plans of the type "if this but not this, then that," we have again the circular paradox of all potential interactions of all possible cell types being preformed or foreseen in the common genome. Neither version of genetic control is required by Elsdale's notion of inherently precise processes. (It is fairly clear in context that he does not mean genetic control in the sense of "characteristic of this species and not of some other." Even if he did, it would not follow that the rules of cellular interaction couldn't vary with conditions.) Indeed, by speaking of constraints being produced by interaction of parts, of internal records of operation being created by morphogenetic processes, he shows how central to his thought is the idea of the development of control, of the dependence of control on past events, of emerging constraints.

3 For more recent descriptions of fibroblast action in morphogenesis, see Lewis (1984). It is, incidentally, kinds of processes and results that are of interest for the present argument, not particular mechanisms. Nor should undue emphasis be placed on random movement. Many processes at this level seem directed by gradients of various types, for example. Though Elsdale's treatment is not recent, his thoughts on the development of biological form seem quite consistent with much contemporary work. What is important is the emergence of order from simpler preexisting order by specifiable mechanisms and the fact that the emerging order is determined in interaction, whether among cells, between cells and chemicals, or whatever. Resulting structures, furthermore, may interact as entities, and the subsequent changes must be explained in terms of those entities, not solely by "information" in their constituents. Løvtrup (1981) makes many of these points in a paper on evolutionary epigenetics. In drawing on Stent's concept of implicit versus explicit genetic information, however, he risks having his ideas assimilated to more conventional notions; what is implicit, after all, is still inherent or essential, albeit not fully or clearly expressed.

4 To apply constructivist principles to the concept of information, rather than accepting genetic and environmental information as givens, is in some sense to extend Piagetian genetic epistemology beyond the realm of logico-mathematical thought whose developmental history he discussed and to apply it to instinct and morphology as well. Though one can find in his work discussions of the transformation of information in development and of genes whose nature is revealed only in their interactions (1971, p. 81), Piaget did not seriously question the concepts of innate structure and instinctive knowledge (1971, pp. 9–18, 66, 227; 1978, pp. 14–22, 33). Why this was so is a complex and intriguing problem, particularly given his willingness to question so much other received wisdom, and given also the power of the conceptual system he elaborated. My own hunch is that Piaget's preoccupation with logic and mathematics as objective, universal, and necessary, and thus different from instinct, which he regarded as contingent and particular (1971, p. 100; 1980, p. 59), kept him from challenging the traditional distinction between the innate and the acquired. He did attempt, using Waddington's concept of genetic assimilation and an idiosyncratic version of the phenocopy, to bridge the gap between them, but because the phenocopy itself was based on the distinction between inherited adaptation and "phenotypic adaptation," the contradictions and ambiguity surrounding these ideas were increased, not resolved (1978, pp. 49, 73–83; 1980). For some general comments on the concept of the phenocopy, see Oyama (1981), and for a discussion of Piaget's use, see Goodwin (1982).

Both Piaget and Lorenz acknowledge Kant's influence on their thinking, and though they disagree on many matters, their concern with the organization of experience by existing structure seems consistent with such influence. Kant was very much aware of the difficulties we encounter in thinking about living order and of the utility (indeed, the necessity, he thought) of using analogies to our own intention to understand it. Though such analogy is a useful guide to our observations of natural phenomena, it cannot, he believed, serve as causal explanation. Since Nature is not intelligent, we can prove her ends only by the "mental jugglery" of reading ends into Nature to make her intelligible (1949, p. 308). (According to McFarland [1970], Kant remained in the grip of the analogy to conscious design and was thus never able to think of undesigned function.) He was also aware that the preformationist and epigenetic theories of his time were both based on preestablished form; he suggested, in fact, that epigenesis, his preferred approach, could also be called "generic preformation" (1949, pp. 338–339).

5 I have always wondered how, since every snowflake is different from every other, but every one is symmetrical, each of the six rays of a flake knew what the others were doing. Packing was the explanation that was once given me;

the chance configuration of particles in the first moments mechanically determined subsequent structure. Though that appeared to explain the variety of flakes, it didn't give a satisfactory explanation of their multifarious intraindividual regularity. Now it seems that snowflake parts are not intelligent and intercommunicating after all, but neither is their form a matter of chain-reaction packing. The microchanges in conditions that occur as the flake moves through the air, along with the properties of water crystals, explain why the six points develop in the same ways: they go through the same conditions at the same moment with the same previously built-up structure, and therefore the same possibilities for additions, thus responding to and creating their own forms (Begley, 1983).

6   The nature of the causal connection between the coenetic variable and the variable by which the system adapts will not be pursued here. I suspect that it is problematic in some ways, but think this approach merits careful attention and energetic attempts to resolve whatever difficulties may exist. Nagel (1979) also believes that the notion of orthogonality is a vulnerable point in Sommerhoff's treatment but does not consider it fatal. Johnston and Turvey (1980) modify the model to include invariant coenetic variables, an alteration that would seem inconsistent with the concept of directive *correlation*. In a personal communication, Johnston says this was done in the interests of elaborating a maximally general model of adaptive organism-environment relations. He does not, however, insist on it very strongly (August 1984).

7   I find it interesting that the innate and the biological these days tend to be identified with the emotions, and usually the baser ones at that. There was a time when ideas, not passions, were inborn, and they were of the true, the good, and the beautiful.

8. Reprise

1   That the levels of analysis, often visualized on the vertical dimension, are often conflated with the back-forward temporal and causal dimensions is probably due in part to the fact that lower levels—chemicals, cells—are the first to be distinguishable as entities in both phylogeny and ontogeny.

2   Though Chomskian linguists have been at pains to show that the multi-leveled complexity of human language makes it qualitatively different from other animal communication, a position with which I fundamentally agree, too facile an assumption has been made by many that the communicatory behavior of other animals is simple and mechanical. For some challenges to this assumption, see Beer (1975a, 1982). One might argue that Chomsky's primary achievement has been the demonstration of the complexity and cre-

ativity of language. Because he lacks, as do we all, a conception of behavioral development that could do justice to these qualities, he declares language to be innate, a tactic that, while emphasizing those central insights, neither helps us understand language nor contributes to a more adequate view of behavioral development in general. An impoverished view of learning requires the innate or something like it to produce the full complement of behavioral abilities. A sufficiently rich view of developmental interactions would not.

3 Is variation in susceptibility to infectious disease, for instance, evidence that disease is actually mental, not physical, and that physical manifestations of such disease are effects, not causes? Andrew Weil, who has had some important and interesting things to say about the way we think about psychoactive drugs, thinks so (1972, pp. 142–144). He admits that the view that illness is caused by lack of psychic harmony is difficult to apply to serious infections in infants. I agree, and suggest this is not a difficulty that will be eliminated, as he claims, when the unconscious mind is better understood, but when the unilinear, oversimplified "straight thinking" he so effectively criticizes is replaced, not by its mirror image (altered states and disease are caused by nonmaterial mind, not by material drugs and genes), but by the interactive systems thinking Weil flirts with but fails, finally, to apply consistently.

For some recent polemics on causation of disease, see exchanges between Lewontin and readers (1983; Lewontin et al., 1983). Complex, pluralistic explanations, hedged with qualifications, lack the impact and immediate gratification of single causes. Every teacher knows the frustration of spending a lecture detailing the wondrous intricacy of a phenomenon, only to have a student ask, "So what really causes it?" (When a colleague described this, I immediately recognized it; hence my generalization to every teacher. It runs a close second to "Do we have to know this?" in the teacher demoralization sweepstakes.) When my son asked me to settle a dispute with a playmate by deciding who had been at fault and I responded with a long-winded exposition of human relations that translated roughly to "Both of you," he rolled his eyes and said plaintively, "Mommy, that's how grown-ups talk, but it's not how kids think." Lest this last example seem frivolous, I will reiterate my conviction that our ideas of causes are closely tied to our ideas of responsibility, and thus to the complicated relations of freedom and intent.

4 See also Ho and Saunders (1982) and Mehler, Morton, and Jusczyk (1984). The latter authors discuss the relationship between psychological and physiological levels in language studies and make the important point that for mapping to be achieved between two levels, adequate theoretical con-

structs must be worked out for both. Even one-to-one mapping with a lower level is not an alternative to doing one's homework at the upper one; indeed, it is a possible prize for good homework. See also Shedletsky (1980) on the problems with hemispheric localization theories of communicative functions.

5 This was a striking and highly effective way to name the phenomenon, and it was an important set of observations to bring to the attention of the field. I admit that my admiration is mixed with regret, insofar as the phrase encourages the stratigraphic instinct-as-deep-and-imperative, learning-as-overlay assumptions that I find so vexing. If, however, when confronted by a trained raccoon that began to "wash" its tokens by rubbing them together instead of completing its routine by placing them in a container, the Brelands (1961) had pointed out that spontaneity is a matter of level and point of view (Hinde, 1974, p. 28), that both species-typical and atypical behavior are prepared and maintained by complex processes, and that cues, movements, and relations can overlap with surprising consequences, maybe no one would have listened to them. In their later book on animal behavior (1966), they present an innovative treatment of these issues without, I believe, referring to instinctive drift at all.

6 When a pattern of perceptions, motives, emotions, behavior, and unpleasant consequences perpetuates itself in spite of the person's unhappiness and lack of a sense of subjective mastery, we may say the person is neurotic. It would be possible to apply the concept of the developmental system here even though the pattern is neither species-typical nor particularly useful in a conventional sense. One of the things neurotic behavior shows us, however, is that usefulness may be assessed on several levels, and that behavior with short-term utility (reducing anxiety, for example) may persist in spite of its long-term costs and in spite of a person's desires.

The example of neurosis is convenient because it shows how an interactive psychological system can show some coherence and continuity without an overall intention to create just that system. One does not, that is, have to postulate a need for pain or for failure in order to gain some insight into such patterns. It is more useful, I think, to try to pay attention to what the person pays attention to and to what interpretations are made; when this is done, what one generally finds is a complex of more conventional motives and ways of perceiving human action (stupid homunculi?) that eventuate in and are sustained by repetitive interactions with others. That such systems can operate in happier ways and in greater harmony with the will is demonstrated by Gruber (1981, 1986), who has elaborated a view of creative work that is neither the effortless unfolding of native gifts nor the impersonal machinations of history but a subtle interplay of emerging aspiration and sustained effort in a particular social context.

7 Despite growing sophistication on these issues, this is emphatically not a straw man. In the context of an otherwise subtle argument on controlled drinking in alcoholics, Stanton Peele declares, "if alcoholism were inbred, mandated by genes, then moderating or reversing it would be impossible" (1984, p. 18). Another example comes from a question asked at a recent convention meeting on personality and disease susceptibility. Is the personality characteristic in question biological, the person wanted to know, or is it alterable?

9. Prospects

1 In doing so, Borges exploits the paradox of making the future present through foreknowledge. As the god simultaneously sees the world's fate and possibly deflects it, so do our perceptions of the inevitable and the possible sometimes alter the very soil from which the future springs. This is one of the reasons I have argued so vehemently against the doctrine of biological fate. Beliefs may fulfill themselves not by virtue of their *truth* but by virtue of their *fixity,* and we are only too ready to disavow responsibility for what we perceive as biologically imposed.

These excerpts from "The God's Script" are from Jorge Luis Borges's *Labyrinths,* copyright © 1962, 1964, by New Directions Publishing Corporation and reprinted by permission of New Directions and Laurence Pollinger, Ltd.

Afterword

1 Keller (1985, 1992) has been particularly acute on these matters, and Nelkin and Lindee (1995) cover much of the ground for the lay audience. See also Neumann-Held (1996) and Rehmann-Sutter (1996).

2 One might object that genetic information *is* material, for it is carried by the complicated structure of the DNA molecule, but if we take this physicality seriously, and I do, then all (!) there is to the idea of genetic information is that marvelously complicated material itself, along with the other (complicated) molecules and (complicated) conditions that are needed for the biochemical processes in question. This includes the mutual responsiveness of the various molecular sites at which the chemicals interact, and,

one might add, the interactive capabilities of wet and dry membranes. No *additional* explanatory role is played by a gene's cognitive or informational powers; in fact, as I have argued, there are prices to pay for these linguistic overgrowths.

3   Inevitability and other causal implications are part of biological determinism, but things are messier than this. As we have seen in the preceding chapters, notions of fundamental nature or meaning are also important: *essence* as well as *incidence*. As is the case with genecentrism in general, these semantic aspects are often more difficult to tease out and counter. If essential meaning were not such a crucial ingredient of this style of biological thinking, informational terms would not be so irresistible, and would not linger so long after all parties had agreed on the causal contingency of the outcomes in question.

4   Levins and Lewontin (1985) use "constructionist," as does Gray (1992). Schaffner (1998) avoids it because of associations with social constructionism. On this count I agree with him, at least insofar as the latter implies treating the social and the biological as alternative sources of form in the time-dishonored manners we have examined. In this Afterword, as in the revision of the book itself, my *constructivist interactionism* appears without the quotation marks that I use for more conventional versions. For a related view, see B. H. Smith's (1997, p. xxi) "constructivist-interactionist account of belief." Tim Ingold (personal communication, July 1998; see also 1995b) assures me that similar disarray surrounds interactionism in anthropology, and the final chapter Howard Becker wrote for the revised edition of his *Outsiders* (1973) not only affords a glimpse of theoretical controversy in sociology but has connections with the debates under discussion here. All of which should alert readers to the minefields they will encounter should they wish to follow my rather casual gestures toward ecological anthropology and symbolic interactionism in Chapter 2.

An example of adherence to dual channel conservatism despite a critical stance toward even more traditional approaches is seen in Sidanius, Cling, and Pratto (1991, pp. 132–133). They inappropriately cite me (along with Levins and Lewontin, 1985) as an advocate of a "biocultural interactionism" whereby "genetically programmed predispositions" like deformities or aggression levels interact with culture. Theirs is a partial "interactionism" that retains genetic and cultural transmission channels even though they "influence" each other. Keeping those separate channels of inheritance is rather like trying to have social integration but maintaining separate entrances. When I was in my early teens, I visited a public bath in Japan. Having disrobed in the women's dressing room I passed through the door to the bath itself and found myself in the common space onto which both dressing room doors opened — one of the shortest and most pointed cultural tutorials I've ever had.

5 Finding out what it means seems simple but can be shockingly hard. Sometimes it is just shocking. An irate behavior geneticist once maintained that everything I had said in a talk was either a truism or simply false. I asked him what he thought had been demonstrated when heritable variation (one meaning of "genetic basis") was found in some analysis. In a presumably unmonitored blurt, he retorted, "That there's something physical involved." I did not have an opportunity to find out what the alternative was, but one needn't be *in* extremis to give this answer; the fact that it is a quite common first response demonstrates how close to the surface the mind-body duality lies. On the meanings of innateness, see Wimsatt (1986, 1999); see Fausto-Sterling (in press) for a view from studies of human sexuality.

6 Or for that matter, any kind of magic bullet, including the drug that does *only one* (totally good) *thing*. See Taylor (1995) for a social-psychological example and Jervis (1997) for a host of others, including many from economics and international relations.

7 They are remarking on Bateson's (1978) review of Dawkins's (1976) *Selfish Gene*. See also my mention in Chapter 7 of rat nests contributing to their own reconstruction in the next generation. Notice that it is the parity move that invites these conjectures by making it *thinkable* that a burrow could be an evolutionary "replicator" (roughly, a self-reproducer, though Sterelny et al. revise this; the model is the selfish gene). One takes the reasoning being used for some kind of developmental resource, in this case genes, and asks whether it can be used for others as well. It is not necessary to accept the concept of selfish replicators to see the conceptual and empirical fertility of the move.

Also pertinent to the question of empirical productivity, and stylistically more congenial to the developmental systems approach, are Gottlieb's (1976) and Bateson's (1983) suggestions for research on the kinds of roles played by various developmental influences, cited in Chapter 8. Links among Tinbergen's (1963) Four Whys, his questions about function, development, evolution, and mechanism, are discussed in Klama (1988). See also Bateson and Martin (1999). We can add the damping out or amplification of change over ontogenetic and transgenerational time (Oyama, 2000, chapter 4; see Jablonka and Lamb, 1995, for cell-level examples).

8 See especially Chapter 8. Many of the sources cited there, such as Johnston (1982), also have valuable discussions of research possibilities. For a more recent summary of some of this work, see Gottlieb (1997). See also Bateson (1987, 1989) on behavior genetic analysis.

9 Elsewhere (Oyama, 2000) I have talked about the evolutionary machismo of writers convinced that "nature-is-nice" sentimentality can be kept at bay by only a properly hardheaded Darwinism. Frans de Waal (1996, pp. 17–19) describes the bias against primatologists' use of words like "reconciliation" or "friendly" to describe chimpanzee behavior. Anthropomorphism is usually the reason for criticism, but as de Waal observes, no such resistance is shown to describing animals as greedy, grudging, spiteful, nepotistic, and murderous.

10 See Griesemer (in press) on frameworks. See also Gray (1987, 1989) on this debate. It may historically be the case that developmental biologists have been interested in universal or highly conserved mechanisms. In addition, however, appropriately informed developmental inquiry could uncover variation in the levels at which conservation occurs. Another of Sterelny's questions has to do, in fact, with homology; I suggest that my view of biological processes makes it quite natural to look for homologous processes and products on different levels, and to expect that they will not always line up neatly with each other. As noted in this book, evolutionists have realized that morphological features can be more highly conserved across phylogenetic time than the genes with which they are associated. The converse can also be true, and this loose-jointedness can be found between other levels as well.

11 Gray's (1988) and Johnston and Gottlieb's (1990) writings could be viewed in this light. Haila (1999) gives a nice discussion of the nature-culture dichotomy in ecology, calling for a developmental perspective on phenomena such as seasonality and succession. He agrees (personal communication, December 1998) that here, as elsewhere, the task is to conceptualize development in a way that makes such efforts something other than the projection of internalist models onto inappropriate objects. In another work (1999b) Haila takes up the challenge of framing research in new ways.

12 See Godfrey-Smith (1996, 1999, in press), Kitcher (in press), and Schaffner (1998). For responses, see Gray (in press), Griffiths and Knight (1998), and Oyama (in press). See also my earlier discussion of causality in systems.

13 This is hardly the place to tackle the thorny question of representation itself. Godfrey-Smith (1996) considers its adequacy in the study of mind to be unresolved, and Johnston (1998) gives some historical perspective. Mention of psychological representations may make psychologists think of James Gibson (1966), whom I discuss briefly in Oyama (1990). Ford and Lerner (1992), whose "developmental systems theory" resembles some of the views presented here but also departs from them in significant ways, seem to take from Gibson just what I decline: support for a cognitivist information-based view of cognition (see below and note 15).

14 See Moss (1992); Neumann-Held (1999), Sarkar (1996), Stent (1986), and Strohman (1997). Other disciplines continue to be represented as well; see Doyle (1997), Nijhout (1990), K. C. Smith (in press).

15 On cognitivism, see Searle (1992), and Varela, Thompson, and Rosch (1991); see also van Gelder (1995), and Clark (1998) on "computationalism." Not all cognitive science is cognitivist in this sense.

For approaches departing from standard cognitivism in various ways, see Caporael (1997), Costall (1995), Hendriks-Jansen (1996), Ingold (1995a, 1996), Shotter (1993), B. C. Smith (1996), B. H. Smith (1997, 1998, February 20), Thelen and Smith (1994), van Gelder (1995), and Varela, Thompson, and Rosch (1991); see also Oyama (1999).

16 The flight from materiality mentioned earlier is sometimes seen in the claim that the *kind* of matter doesn't matter, that any material substrate in the right relations can produce, have, or be a mind. Searle, like Chalmers, insists that we must deal with consciousness, but he denies that the language of computation holds the key. Central as it is to these debates, computation itself remains somewhat mysterious. There seems to be some stage magic involved in the "transduction" by which the physical energies of sensory stimulation become the information of which mental representations are made; make a quick enough costume change, from body to mind, say, and no one notices you've switched plays as well. B. C. Smith (1996) argues that disputants on both sides of the argument over computational minds labor under multiple misconceptions about computation, computers, and much else besides.

References

Alberch, P. (1982). Developmental constraints in evolutionary processes. In J. T. Bonner (Ed.), *Evolution and development* (313–332). Berlin: Springer-Verlag.

Alberts, B., Bray, D., Lewis, J., Raff, M., Roberts, K., & Watson, J. D. (1983). *Molecular biology of the cell.* New York: Garland.

Allen, G. E. (1979). Naturalists and experimentalists: The genotype and the phenotype. In W. Coleman & C. Limoges (Eds.), *Studies in history of biology* (Vol. 3) (179–210). Baltimore: Johns Hopkins University Press.

Anastasi, A. (1958). Heredity, environment and the question "how?" *Psychological Review, 65,* 197–208.

Ayala, F. J. (1972). The autonomy of biology as a natural science. In A. D. Breck & W. Yourgrau (Eds.), *Biology, history, and natural philosophy* (1–16). New York: Plenum.

Barash, D. P. (1977). *Sociobiology and behavior.* New York: Elsevier.

Bateson, G. (1972). *Steps toward an ecology of mind.* New York: Ballantine Books.

Bateson, P. P. G. (1976). Rules and reciprocity in development. In P. P. G. Bateson & R. A. Hinde (Eds.), *Growing points in ethology* (401–421). Cambridge: Cambridge University Press.

Bateson, P. P. G. (1978). [Review of *The Selfish Gene*]. *Animal Behaviour, 26,* 316–318.

Bateson, P. P. G. (1982). Behavioural development and evolutionary processes. In King's College Sociobiology Group (Eds.), *Current problems in sociobiology* (133–151). Cambridge: Cambridge University Press.

Bateson, P. [P. G.] (1983). Genes, environment and the development of behaviour. In T. R. Halliday & P. J. B. Slater (Eds.), *Animal behaviour: Genetics and development* (52–81). Oxford: Blackwell.

Bateson, P. [P. G.] (1987). Biological approaches to the study of behavioural development. *International Journal of Behavioral Development, 10,* 1–22.

Bateson, P. [P. G.] (1989). Additive models may mislead. *International Journal of Behavioral Development, 12,* 407–411.

Bateson, P. P. G., & Klopfer, P. H. (Eds.). (1981). *Perspectives in ethology* (Vol. 4). New York: Plenum.

Bateson, P. [P. G.], & Martin, P. (1999). *Design for a life.* London: Jonathan Cape.

Beach, F. A. (1955). The descent of instinct. *Psychological Review, 62,* 401–410.

Becker, S. (1973). *Outsiders: Studies in the sociology of deviance* (rev. ed.). New York: Free Press.

Beckner, M. D. (1967). Vitalism. In P. Edwards (Ed.), *The encyclopedia of philosophy* (Vol. 8) (253–256). New York: Macmillan and Free Press.

Beer, C. G. (1975a). Multiple functions and gull displays. In G. Baerends, C. Beer, & A. Manning (Eds.), *Function and evolution in behaviour* (16–55). Oxford: Oxford University Press.

Beer, C. G. (1975b). Was Professor Lehrman an ethologist? *Animal Behavior, 23,* 957–964.

Beer, C. G. (1982). Study of vertebrate communication—its cognitive implications. In D. R. Griffin (Ed.), *Animal mind–human mind* (251–268). Berlin: Springer-Verlag.

Beer, C. G. (1983a). Darwin, instinct, and ethology. *Journal of the History of the Behavioral Sciences, 19,* 68–80.

Beer, C. [G.] (1983b). Minds and machines: Motive for metaphor? A response to Boden. *New Ideas in Psychology, 1,* 117–122.

Begley, S., with Carey, J. (December 26, 1983). *Newsweek,* 64.

Benedict, R. (1934). *Patterns of culture.* Boston: Houghton Mifflin.

Bertalanffy, L. von. (1967). *Robots, men and minds.* New York: Braziller.

Bertalanffy, L. von. (1972). The model of open systems: Beyond molecular biology. In A. D. Breck & W. Yourgrau (Eds.), *Biology, history, and natural philosophy* (217–230). New York: Plenum.

Block, N. (1979). A confusion about innateness. *Behavioral and Brain Sciences, 2,* 27–29.

Bonner, J. T. (1974). *On development.* Cambridge: Harvard University Press.

Bowers, K. S. (1973). Situationism in psychology: An analysis and a critique. *Psychological Review, 80,* 307–336.

Bowers, M. D. (1980). Unpalatability as a defense strategy of Euphydryas phaeton (*Lepidoptera: nymphaeidae*). *Evolution, 34,* 586–600.

Breland, K., & Breland, M. (1961). The misbehavior of organisms. *American Psychologist, 16,* 681–684.

Breland, K., & Breland, M. (1966). *Animal behavior.* New York: Macmillan.

Brim, O. G. Jr., & Kagan, J. (1980). Constancy and change: A view of the issues. In O. G. Brim Jr., & J. Kagan (Eds.), *Constancy and change in human development* (1–25). Cambridge: Harvard University Press.

Brody, J. E. (January 17, 1984). Self-blame held to be important in victims' recovery. *New York Times,* C1, C6.

Brower, L. P., & Glazier, S. C. (1975). Localization of heart poisons in the monarch butterfly. *Science, 188,* 19–25.

Burian, R. M. (1978). A methodological critique of sociobiology. In A. L. Caplan (Ed.), *The sociobiology debate* (376–396). New York: Harper and Row.

Burian, R. M. (1981–1982). Human sociobiology and genetic determinism. *Philosophical Forum, 13* (2–3), 43–66.

Burks, A. W. (1966). Editor's introduction. In J. von Neumann, *Theory of self-reproducing automata* (1–28). Edited and completed by A. W. Burks. Urbana: University of Illinois Press.

Campbell, J. (1982). *Grammatical man.* New York: Simon and Schuster.

Caplan, A. L. (Ed.). (1978). *The sociobiology debate.* New York: Harper and Row.

Caplan, A. L. (1981–1982). Say it just ain't so: Adaptational stories and socio-biological explanations of social behavior. *Philosophical Forum, 13* (2–3), 144–160.

Caporael, L. R. (1997). The evolution of truly social cognition: The core configuration model. *Personality and Social Psychology Review, 1,* 276–298.

Cartmill, M. (November 1983). "Four legs good, two legs bad." *Natural History,* 64–79.

Chalmers, D. J. (1996). *The conscious mind: In search of a fundamental theory.* Oxford: Oxford University Press.

Chomsky, N. (1972). *Problems of knowledge and freedom: The Russell lectures.* New York: Random House/Vintage Books.

Clark, A. (1998). Time and mind. *Journal of Philosophy, 95,* 354–376.

Clutter, M. E. (1978). *Dormancy and developmental arrest.* New York: Academic Press.

Conrad, M. (1972). The importance of molecular hierarchy in information processing. In C. H. Waddington (Ed.), *Towards a theoretical biology* (Vol. 4) (222–228). Edinburgh: Edinburgh University Press.

Costall, A. (1995). Socializing affordances. *Theory and Psychology, 5,* 467–482.

Dawkins, R. (1976). *The selfish gene.* Oxford: Oxford University Press.

Dawkins, R. (1982). *The extended phenotype: The gene as the unit of selection.* San Francisco: Freeman.

Dawkins, R. (1984). Replicator selection and the extended phenotype. In E. Sober (Ed.), *Conceptual issues in evolutionary biology: An anthology* (125–141). Cambridge: MIT Press/Bradford Books.

Deacon, T. W. (1997). *The symbolic species: The co-evolution of language and the brain.* New York: Norton.

de Beer, G. (1958). *Embryos and ancestors.* Oxford: Oxford University Press/ Clarendon Press.

Dennett, D. C. (1995). *Darwin's dangerous idea.* New York: Simon and Schuster.

Dinnerstein, D. (1977). *The mermaid and the minotaur.* New York: Harper Colophon Books.

Doolittle, W. F. (1982). Selfish DNA after fourteen months. In G. A. Dover & R. B. Flavell (Eds.), *Genome evolution* (3–28). New York: Academic Press.

Doyle, R. M. (1997). *On beyond living: Rhetorical transformations of the life sciences.* Stanford: Stanford University Press.

Dretske, F. I. (1983). *Knowledge and the flow of information.* Cambridge: MIT Press/Bradford Books.

Dunbar, R. I. M. (1982). Adaptation, fitness and the evolutionary tautology. In King's College Sociobiology Group (Eds.), *Current problems in sociobiology* (9–28). Cambridge: Cambridge University Press.

Eibl-Eibesfeldt, I. (1979). Human ethology: Concepts and implications for the sciences of man. *Behavioral and Brain Sciences, 2,* 1–57.

Eigen, M. (1967). Dynamic aspects of information transfer and reaction control in biomolecular systems. In G. C. Quarton, T. Melnechuk, & F. O. Schmitt (Eds.), *The neurosciences: A study program* (130–142). New York: Rockefeller University Press.

Eigen, M., & DeMaeyer, L. C. M. (1966). Summary of two NRP work sessions on information storage and processing in biomolecular systems. In F. O. Schmitt & T. Melnechuk (Eds.), *Neurosciences research symposium summaries* (244–266). Cambridge: MIT Press.

Elman, J. L., Bates, E. A., Johnson, M. H., Karmiloff-Smith, A. Parisi, D., & Plunkett, K. (1996). *Rethinking innateness: A connectionist perspective on development.* Cambridge: MIT Press/Bradford Books.

Elsdale, T. (1972). Pattern formation in fibroblast cultures, an inherently precise morphogenetic process. In C. H. Waddington (Ed.), *Towards a theoretical biology* (Vol. 4) (259–275). Edinburgh: Edinburgh University Press.

Fausto-Sterling, A. (in press). *Sexing the body: Gender politics and the construction of human sexuality.* New York: Basic Books.

Fernberger, S. W. (1929). Unlearned behavior of the albino rat. *American Journal of Psychology, 41,* 343–344.

Fishbein, H. D. (1976). *Evolution, development, and children's learning.* Santa Monica, Calif.: Goodyear.

Fodor, J. (1980). Fixation of belief and concept acquisition. In M. Piattelli-Palmarini (Ed.), *Language and learning* (142–149, and discussion elsewhere in volume). Cambridge: Harvard University Press.

Ford, D. H., & Lerner, R. M. (1992). *Developmental systems theory: An integrative approach.* Newbury Park, Calif.: Sage.

Foster, D. B. (1975). *Intelligent universe: A cybernetic philosophy.* New York: Putnam.

Fredrick, J. F. (Ed.). (1981). *Origins and evolution of eukaryotic intracellular organelles.* Annals of the New York Academy of Sciences (Vol. 361). New York: New York Academy of Sciences.

Gasking, E. (1967). *Investigations into generation: 1651–1828.* Baltimore: Johns Hopkins University Press.

Gatlin, L. L. (1972). *Information theory and the living system.* New York: Columbia University Press.

Geertz, C. (1973). The impact of the concept of culture on the concept of man. In C. Geertz (Ed.), *The interpretation of cultures* (33–54). New York: Basic Books.

Gesell, A. (1945). The embryology of behavior. Westport, Conn.: Greenwood Press.

Gesell, A. (1954). The ontogenesis of infant behavior. In L. Carmichael (Ed.), *Manual of child psychology* (2nd ed.) (335–373). New York: Wiley.

Gibson, J. J. (1966). *The senses considered as perceptual systems.* Boston: Houghton Mifflin.

Godfrey-Smith, P. (1996). *Complexity and the function of mind in nature.* Cambridge: Cambridge University Press.

Godfrey-Smith, P. (1999). Genes and codes: Lessons from the philosophy of mind? In V. Hardcastle (Ed.), *Biology meets psychology: Constraints, connections, conjectures* (305–331). Cambridge: MIT Press.

Godfrey-Smith, P. (in press). Explanatory symmetries, preformation and developmental systems theory. *Philosophy of Science.*

Goodwin, B. C. (1970). Biological stability. In C. H. Waddington (Ed.), *Towards a theoretical biology* (Vol. 3) (1–17). Chicago: Aldine.

Goodwin, B. C. (1972). Biology and meaning. In C. H. Waddington (Ed.), *Towards a theoretical biology* (Vol. 4) (259–275). Edinburgh: Edinburgh University Press.

Goodwin, B. C. (1982). Genetic epistemology and constructionist biology. *Revue internationale de philosophie, 142–143,* 527–548.

Goodwin, B. C., & Trainor, L. E. H. (1983). The ontogeny and phylogeny of the pentadactyl limb. In B. C. Goodwin, N. Holder, & C. G. Wylie (Eds.), *Development and evolution* (75–98). Cambridge: Cambridge University Press.

Gordon, D. M. (1991). Behavioral flexibility and the foraging ecology of seed-eating ants. *American Naturalist, 138,* 379–411.

Gottlieb, G. (1975). Development of species identification in ducklings: II. Experiential prevention of perceptual deficit caused by embryonic auditory deprivation. *Journal of Comparative and Physiological Psychology, 89,* 675–684.

Gottlieb, G. (1976). Conceptions of prenatal development: Behavioral embryology. *Psychological Review, 83,* 215–234.

Gottlieb, G. (1997). *Synthesizing nature-nurture: Prenatal roots of instinctive behavior.* Mahwah, N.J.: Erlbaum.

Gould, S. J. (1977). *Ontogeny and phylogeny.* Cambridge: Harvard University Press/Belknap Press.

Gould, S. J. (1981). *The mismeasure of man.* New York: Norton.

Gould, S. J. (1982). Darwinism and the expansion of evolutionary theory. *Science, 216,* 380–387.

Gould, S. J. (August 12, 1984). Similarities between the sexes. *New York Times Book Review,* 7.

Gray, R. D. (1987). Beyond labels and binary oppositions: What can be learnt from the nature/nurture dispute? *Rivista di Biologia/Biological Forum, 80,* 192–196.

Gray, R. D. (1988). Metaphors and methods: Behavioural ecology, panbiogeography and the evolving synthesis. In M.-W. Ho & S. W. Fox (Eds.), *Evolutionary processes and metaphors* (209–242). London: Wiley.

Gray, R. D. (1989). Oppositions in panbiogeography: Can the conflicts between selection, constraint, ecology, and history be resolved? *New Zealand Journal of Zoology, 16,* 787–806.

Gray, R. D. (1992). Death of the gene: Developmental systems strike back. In P. [E.] Griffiths (Ed.), *Trees of life: Essays in philosophy of biology* (165–209). Dordrecht, Netherlands: Kluwer Academic.

Gray, R. D. (in press). Selfish genes or developmental systems? Evolution without replicators and vehicles. In R. Singh, K. Krimbas, D. Paul, & J. Beatty (Eds.), *Thinking about evolution: Historical, philosophical and political perspectives.* Cambridge: Cambridge University Press.

Greenberg, G., & Tobach, E. (Eds.). (1984). *Behavioral evolution and integrative levels.* Hillsdale, N.J.: Erlbaum.

Griesemer, J. (in press). Development, culture, and the units of inheritance. *Philosophy of Science.*

Griffin, D. R. (Ed.) (1982). *Animal mind–human mind.* Berlin: Springer-Verlag.

Griffin, D. R. (1984). Animal thinking. Cambridge: Harvard University Press.

Griffiths, P. E., & Gray, R. D. (1994). Developmental systems and evolutionary explanation. *Journal of Philosophy, 91,* 277–304.

Griffiths, P. E., & Gray, R. D. (1997). Replicator II: Judgement day. *Biology and Philosophy, 12,* 471–492.

Griffiths, P. E., & Knight, R. D. (1998). What is the developmentalist challenge? *Philosophy of Science, 65,* 253–258.

Grmek, M. D. (1972). A survey of the mechanical interpretations of life from the Greek atomists to the followers of Descartes. In A. D. Breck & W. Yourgrau (Eds.), *Biology, history, and natural philosophy* (181–195). New York: Plenum.

Grobstein, C. (1974). *The strategy of life* (2nd ed.). San Francisco: Freeman.

Gruber, H. E. (1977). The fortunes of a basic Darwinian idea: Chance. In R. W.

Rieber & K. Salzinger (Eds.), *The roots of American psychology: Historical influences and implications for the future. Annals of the New York Academy of Sciences* (Vol. 291) (233–245). New York: New York Academy of Sciences.

Gruber, H. [E.] (1980). Afterword. In D. H. Feldman, *Beyond universals in cognitive development* (175–180). Norwood, N. J.: Ablex.

Gruber, H. E. (1981). *Darwin on man, a psychological study of scientific creativity* (2nd ed.). Chicago: University of Chicago Press.

Gruber, H. E. (1983). History and creative work: From the most ordinary to the most exalted. *Journal of the History of the Behavioral Sciences, 19,* 4–14.

Gruber, H. E. (1986). The self-construction of the extraordinary. In R. Sternberg & J. Davidson (Eds.), *Conceptions of giftedness* (247–263). Cambridge: Cambridge University Press.

Haila, Y. (1999a). "Biodiversity" and the nature-culture divide: Conflicting tendencies. *Biodiversity and Conservation, 8,* 165–181.

Haila, Y. (1999b). Socioecologies. *Ecography, 22,* 337–348.

Hailman, J. P. (1982). Ontogeny: Toward a general theoretical framework for ethology. In P. P. G. Bateson & P. H. Klopfer (Eds.), *Perspectives in ethology* (Vol. 5) (133–189). New York: Plenum.

Hamburger, V. (1980). Embryology and the modern synthesis in evolutionary theory. In E. Mayr & W. B. Provine (Eds.), *The evolutionary synthesis* (97–112). Cambridge: Harvard University Press.

Harris, M. (July/August 1983). Margaret and the giant-killer. *Sciences,* 18–21.

Heider, F. (1958). *The psychology of interpersonal relations.* New York: Wiley.

Hendriks-Jansen, H. (1996). *Catching ourselves in the act.* Cambridge: MIT Press/Bradford Books.

Hinde, R. A. (1968). Dichotomies in the study of development. In J. M. Thoday & A. S. Parkes (Eds.), *Genetic and environmental influences on behaviour* (3–14). New York: Plenum.

Hinde, R. A. (1970). *Animal behaviour* (2nd ed.). New York: McGraw-Hill.

Hinde, R. A. (1974). *Biological bases of human social behavior.* New York: McGraw-Hill.

Hinde, R. A., & Bateson. P. [P. G.]. (1984). Discontinuities versus continuities in behavioural development and the neglect of process. *International Journal of Behavioral Development, 7,* 129–143.

Hinde, R. A., & Stevenson-Hinde, J. (1973). *Constraints on learning.* New York: Academic Press.

Ho, M. W., & Saunders, P. T. (1979). Beyond neo-Darwinism—an epigenetic approach to evolution. *Journal of Theoretical Biology, 78,* 573–591.

Ho, M. W., & Saunders, P. T. (1982). The epigenetic approach to the evolution of organisms—with notes on its relevance to social and cultural evolution.

In H. C. Plotkin (Ed.), *Learning, development, and culture* (343–360). New York: Wiley.

Hofer, M. A. (1981). Parental contributions to the development of their offspring. In D. J. Gubernick & P. H. Klopfer (Eds.), *Parental care in mammals* (77–115). New York: Plenum.

Hofstadter, D. R. (1980). *Gödel, Escher, Bach: An eternal golden braid.* New York: Random House/Vintage Books.

Hull, D. L. (1984). A matter of individuality. In E. Sober (Ed.), *Conceptual issues in evolutionary biology* (623–645). Cambridge: MIT Press.

Hyland, M. E. (1984). Interactionism and the person × situation debate. In J. R. Royce & L. P. Mos (Eds.), *Annals of theoretical psychology* (Vol. 2) (303–328). New York: Plenum.

Immelmann, K., Barlow, G. W., Petrinovich, L., & Main, M. (1981). General introduction. In K. Immelmann, G. W. Barlow, L. Petrinovich, & M. Main (Eds.), *Behavioral development* (1–18). Cambridge: Cambridge University Press.

Ingold, T. (1995a). Building, dwelling, living: How animals and people make themselves at home in the world. In M. Strathern (Ed.), *Shifting contexts: Transformations in anthropological knowledge* (57–70). London: Routledge.

Ingold, T. (1995b). 'People like us': The concept of the anatomically modern human. *Cultural Dynamics, 7,* 187–214.

Ingold, T. (1996). Culture, perception and cognition. In J. Haworth (Ed.), *Psychological research: Innovative methods and strategies* (99–119). London: Routledge.

Jablonka, E., & Lamb, M. J. (1995). *Epigenetic inheritance and evolution: The Lamarckian dimension.* Oxford: Oxford University Press.

Jacob, F. (1973). *The logic of life: A history of heredity* (B. E. Spillman, Trans.). New York: Pantheon Books. (Original work published 1970.)

Jacob, F. (1982). *The possible and the actual.* Seattle: University of Washington Press.

Jacobson, M. (1974). A plenitude of neurons. In G. Gottlieb (Ed.), *Aspects of neurogenesis* (Vol. 2) (151–166). New York: Academic Press.

Jacobson, M. (1978). *Developmental neurobiology* (2nd ed.). New York: Plenum.

Jerne, N. K. (1967). Antibodies and learning: Selection versus instruction. In G. C. Quarton, T. Melnechuk, & F. O. Schmitt (Eds.), *The neurosciences: A study program* (200–205). New York: Rockefeller University Press.

Jervis, R. (1997). *System effects: Complexity in political and social life.* Princeton: Princeton University Press.

Johnston, T. D. (1982). Learning and the evolution of developmental systems. In H. C. Plotkin (Ed.), *Learning, development, and culture* (411–442). New York: Wiley.

Johnston, T. D. (1998). Comment on Miller. In C. Dent-Read & P. Zukow-Goldring (Eds.), *Evolving explanations of development* (509–513). Washington, D.C.: American Psychological Association.

Johnston, T. D., & Gottlieb, G. (1990). Neophenogenesis: A developmental theory of phenotypic evolution. *Journal of Theoretical Biology, 147,* 471–495.

Johnston, T. D., & Turvey, M. T. (1980). A sketch of an ecological metatheory for theories of learning. In G. H. Bower (Ed.), *The psychology of learning and motivation* (Vol. 14) (147–205). New York: Academic Press.

Kant, I. (1949). Critique of judgment (J. C. Meredith, Trans.). In C. J. Friedrich (Ed.), *The philosophy of Kant* (265–364). New York: Random House/Modern Library.

Keegan, R. T., & Gruber, H. E. (1983). Love, death, and continuity in Darwin's thinking. *Journal of the History of the Behavioral Sciences, 19,* 15–30.

Keller, E. F. (1985). *Reflections on gender and science.* New Haven: Yale University Press.

Keller, E. F. (1992). *Secrets of life, secrets of death.* New York: Routledge.

King, J. A. (1968). Species specificity and early experience. In G. Newton & S. Levine (Eds.), *Early experience and behavior* (42–64). Springfield, Ill.: Thomas.

Kitchener, R. F. (1978). Epigenesis: The role of biological models in developmental psychology. *Human Development, 21,* 141–160.

Kitchener, R. F. (1980). Predictive versus probabilistic epigenesis. *Human Development, 23,* 73–76.

Kitcher, P. (in press). Battling the undead: How (and how not) to resist genetic determinism. In R. Singh, K. Krimbas, D. Paul, & J. Beatty (Eds.), *Thinking about evolution: Historical, philosophical and political perspectives.* Cambridge: Cambridge University Press.

Klama, J. (1988). *The myth of the beast within: Aggression revisited.* London: Longman Group (Published in the United States by Wiley, as *Aggression: The myth of the beast within.*)

Klopfer, P. [H.] (1969). Instincts and chromosomes: What is an "innate" act? *American Naturalist, 103,* 556–560.

Klopfer, P. H. (1973). *On behavior: Instinct is a Cheshire cat.* Philadelphia: Lippincott.

Klopfer, P. H. (1981). Origins of parental care. In D. J. Gubernick & P. H. Klopfer (Eds.), *Parental care in mammals* (1–12). New York: Plenum.

Koestler, A. (1967). *The ghost in the machine.* New York: Macmillan.

Kolata, G. (1984). New clues to gene regulation. *Science, 224,* 588–589.

Konner, M. (1982). *The tangled wing: Biological constraints on the human spirit.* New York: Holt, Rinehart and Winston.

Kummer, H. (1971). *Primate societies.* Chicago: Aldine Atherton.

Kummer, H. (1980). Analogs of morality among nonhuman primates. In G. S. Stent (Ed.), *Morality as a biological phenomenon* (31–47). Berkeley: University of California Press.

Lehrman, D. S. (1970). Semantic and conceptual issues in the nature-nurture problem. In L. R. Aronson, E. Tobach, D. S. Lehrman, & J. S. Rosenblatt (Eds.), *Development and evolution of behavior* (17–52). San Francisco: Freeman.

Leighton, T., & Loomis, W. F. (1980). Introduction. In T. Leighton & W. F. Loomis (Eds.), *The molecular genetics of development* (xiii–xxiii). New York: Academic Press.

Lerner, R. M. (1980). Concepts of epigenesis: Descriptive and explanatory issues. *Human Development, 23,* 63–72.

Lerner, R. M., & Busch-Rossnagel, N. A. (Eds.). (1981). *Individuals as producers of their development.* New York: Academic Press.

Lerner, R. M., Hultsch, D. F., & Dixon, R. A. (1983). Contextualism and the character of developmental psychology in the 1970s. In J. W. Dauben & V. S. Sexton (Eds.), *History and philosophy of science. Annals of the New York Academy of Sciences* (Vol. 412) (101–128). New York: New York Academy of Sciences.

Levins, R., & Lewontin, R. (1985). *The dialectical biologist.* Cambridge: Harvard University Press.

Lewin, R. (1984). Why is development so illogical? *Science, 224,* 1327–1329.

Lewis, J. (1984). Morphogenesis by fibroblast traction. *Nature, 307,* 413–414.

Lewontin, R. C. (September 1978). Adaptation. *Scientific American,* 212–230.

Lewontin, R. C. (1982). Organism and environment. In H. C. Plotkin (Ed.), *Learning, development, and culture* (151–170). New York: Wiley.

Lewontin, R. C. (January 20, 1983). The corpse in the elevator. *New York Review of Books,* 34–37.

Lewontin, R. C., and others. (April 28, 1983). "The corpse in the elevator": An exchange. *New York Review of Books,* 42–43.

Lewontin, R. C., Rose, S., & Kamin, L. J. (1984). *Not in our genes.* New York: Pantheon.

Long, C. H. (1963). *Alpha: The myths of creation.* New York: Braziller.

Lorenz, K. (1965). *Evolution and modification of behavior.* Chicago: University of Chicago Press.

Lorenz, K. (1977). *Behind the mirror* (R. Taylor, Trans.). New York: Harcourt Brace Jovanovich. (Original work published 1973.)

Løvtrup, S. (1981). Introduction to evolutionary epigenetics. In G. G. E. Scudder & J. L. Reveal (Eds.), *Evolution today. Proceedings of the 2nd International Congress of Systematic and Evolutionary Biology* (131–144). Pittsburgh, Pa.: Hunt Institute for Botanical Documentation, Carnegie-Mellon University.

Lumsden, C. J., & Wilson, E. O. (1981). *Genes, mind, and culture.* Cambridge: Harvard University Press.

MacWilliams, H. K., & Bonner, J. T. (1979). The prestalk-prespore pattern in cellular slime molds. *Differentiation, 14,* 1–22.

Margulis, L. (1974). Symbiosis and evolution. In E. O. Wilson (Comp.), *Ecology, evolution, and population biology: Readings from* Scientific American (179–188). San Francisco: Freeman.

Margulis, L. (1981). *Symbiosis in cell evolution: Life and its environment on the early Earth.* San Francisco: Freeman.

Mattern, R. (1978). Altruism, ethics, and sociobiology. In A. L. Caplan (Ed.), *The sociobiology debate* (462–475). New York: Harper and Row.

May, R. M. (September 1978). The evolution of ecological systems. *Scientific American,* 160–175.

Mayr, E. (1976). *Evolution and the diversity of life.* Cambridge: Harvard University Press/Belknap Press.

Mayr, E. (1982). *The growth of biological thought.* Cambridge: Harvard University Press/Belknap Press.

McClearn, G. E., & DeFries, J. C. (1973). *Introduction to behavioral genetics.* San Francisco: Freeman.

McFarland, J. D. (1970). *Kant's concept of teleology.* Edinburgh: Edinburgh University Press.

Mehler, J., Morton, J., & Jusczyk, P. W. (1984). On reducing language to biology. *Cognitive Neuropsychology, 1,* 83–116.

Mercer, E. H. (1981). *The foundations of biological theory.* New York: Wiley.

Merchant, C. (1982). Isis' consciousness raised. *Isis, 73,* 398–409.

Midgley, M. (1979). Gene-juggling. *Philosophy, 54,* 439–458.

Midgley, M. (1980). *Beast and man: The roots of human nature.* New York: New American Library.

Miller, G. A. (1973). Psychology and communication. In G. A. Miller (Ed.), *Communication, language, and meaning* (3–12). New York: Basic Books.

Miller, G. A., Galanter, E., & Pribram, K. (1960). *Plans and the structure of behavior.* New York: Holt, Rinehart and Winston.

Monod, J. (1971). *Chance and necessity* (A. Wainhouse, Trans.). New York: Knopf.

Montague, M. F. A. (1966). Constitutional and prenatal factors in infant and child health. In M. J. Senn (Ed.), *Symposium on the healthy personality* (148–210). New York: Josiah Macy Jr. Foundation.

Montalenti, G. (1974). From Aristotle to Democritus via Darwin. In F. J. Ayala & T. Dobzhansky (Eds.), *Studies in the philosophy of biology* (3–19). Berkeley: University of California Press.

Moss, L. (1992). A kernel of truth? On the reality of the genetic program. In D. L.

Hull, M. Forbes, & K. Okruhlik (Eds.), *Philosophy of Science Association proceedings, 1,* 335–348.

Munn, N. L. (1965). *The evolution and growth of human behavior* (2nd ed.). Boston: Houghton Mifflin.

Murphy, C. (November 1997). DNA fatigue: Worn out by a nucleotidal wave. *Atlantic Monthly, 28,* 30.

Nagel, E. (1979). Teleology revisited. In E. Nagel (Ed.), *Teleology revisited* (275–316). New York: Columbia University Press.

Nelkin, D., & Lindee, M. S. (1995). *The DNA mystique: The gene as a cultural icon.* New York: Freeman.

Netting, R. M. (1978). *Cultural ecology.* Menlo Park, Calif.: Cummings.

Neumann, J. von. (1966). *Theory of self-reproducing automata* (Edited and completed by A. W. Burks). Urbana: University of Illinois Press.

Neumann-Held, E. M. (1996). Die moderne Biologie auf der Suche nach dem "Heiligen Gral." [Modern biology in search of the "Holy Grail."] In T. Fischer & R. Seising (Eds.), *Wissenschaft und Oeffentlichkeit* (135–159). Frankfurt/M: Peter Lang.

Neumann-Held, E. M. (1999). The gene is dead—long live the gene! Conceptualizing genes the constructionist way. In P. Koslowski (Ed.), *Sociobiology and bioeconomics. The theory of evolution in biological and economic theory* (105–137). Studies in Economic Ethics and Philosophy (Vol. 20). Berlin: Springer-Verlag.

Newman, S. A. (January/February 1984). Vertebrate bones and violin tones. *Sciences,* 38–43.

Nijhout, H. F. (1990). Metaphors and the role of genes in development. *BioEssays, 12,* 441–446.

Nitecki, M. H. (Ed.). (1983). *Coevolution.* Chicago: University of Chicago Press.

Niven, L., & Pournelle, J. (1975). *The mote in God's eye.* New York: Pocket Books.

Oppenheim, R. W. (1980). Metamorphosis and adaptation in the behavior of developing organisms. *Developmental Psychobiology, 13,* 353–356.

Oppenheim, R. W. (1982). Preformation and epigenesis in the origins of the nervous system and behavior: Issues, concepts, and their history. In P. P. G. Bateson & P. H. Klopfer (Eds.), *Perspectives in ethology* (Vol. 5) (1–100). New York: Plenum.

Oyama, S. (1979). The concept of the sensitive period in developmental studies. *Merrill-Palmer Quarterly, 25,* 83–103.

Oyama, S. (1981). What does the phenocopy copy? *Psychological Reports, 48,* 571–581.

Oyama, S. (1982). A reformulation of the concept of maturation. In P. P. G. Bateson & P. H. Klopfer (Eds.), *Perspectives in ethology* (Vol. 5) (101–131). New York: Plenum.

Oyama, S. (1990). Commentary. The idea of innateness: Effects on language and communication research. *Developmental Psychobiology, 23,* 741–747.

Oyama, S. (1999). Locating development, locating developmental systems. In E. K. Scholnick, K. Nelson, S. A. Gelman, & P. H. Miller (Eds.), *Conceptual development: Piaget's legacy* (185–208). Hillsdale, N.J.: Erlbaum.

Oyama, S. (2000). *Evolution's eye: A systems view of the biology-culture divide.* Durham, N.C.: Duke University Press.

Oyama, S. (in press-a). Causal democracy and causal contributions in DST. *Philosophy of Science.*

Oyama, S. (in press-b). Terms in tension: What do you do when all the good words are taken? In S. Oyama, P. E. Griffiths, & R. D. Gray (Eds.), *Cycles of contingency: Developmental systems and evolution.* Cambridge: MIT Press/ Bradford Books.

Oyama, S., Griffiths, P. E., & Gray, R. D. (Eds.). (in press). *Cycles of contingency: Developmental systems and evolution.* Cambridge: MIT Press/Bradford Books.

Paigen, K. (1980). Temporal genes and other developmental regulators in mammals. In T. Leighton & W. F. Loomis (Eds.), *The molecular genetics of development* (419–470). New York: Academic Press.

Pattee, H. H. (1972). Laws and constraints, symbols and languages. In C. H. Waddington (Ed.), *Towards a theoretical biology* (Vol. 4) (248–257). Edinburgh: Edinburgh University Press.

Peele, S. (March/April 1984). The new prohibitionists. *Sciences,* 14–19.

Peterson, S. A., & Somit, A. (1978). Sociobiology and politics. In A. L. Caplan (Ed.), *The sociobiology debate* (449–461). New York: Harper and Row.

Piaget, J. (1971). *Biology and knowledge* (B. Walsh, Trans.). Chicago: University of Chicago Press. (Original work published 1967.)

Piaget, J. (1978). *Behavior and evolution* (D. Nicholson-Smith, Trans.). New York: Pantheon Books. (Original work published 1976.)

Piaget, J. (1980). Introductory remarks. In M. Piattelli-Palmarini (Ed.), *Language and learning* (57–61). Cambridge: Harvard University Press.

Pittendrigh, C. (1958). Adaptation, natural selection, and behavior. In A. Roe & G. G. Simpson (Eds.), *Behavior and evolution* (390–416). New Haven: Yale University Press.

Plomin, R. (1981). Ethological behavioral genetics and development. In K. Immelmann, G. W. Barlow, L. Petrinovich, & M. Main (Eds.), *Behavioral development* (252–276). Cambridge: Cambridge University Press.

Plomin, R., & Rowe, D. C. (1978). Genes, environment, and development of temperament in young human twins. In G. M. Burghardt & M. Bekoff (Eds.),

The development of behavior: Comparative and evolutionary aspects (279–296). New York: Garland Press.

Pribram, K. H. (1980). The role of analogy in transcending limits in the brain sciences. *Daedalus, 109* (2), 19–38.

Prigogine, I., & Stengers, I. (1984). *Order out of chaos.* New York: Bantam Books. (Original work published 1979.)

Raff, R. A., & Kaufman, T. C. (1983). *Embryos, genes, and evolution: The developmental-genetic basis of evolutionary change.* New York: Macmillan.

Ransom, R. (1981). *Computers and embryos.* New York: Wiley.

Ravin, A. W. (1977). The gene as catalyst; the gene as organism. In W. Coleman & C. Limoges (Eds.), *Studies in history of biology* (1–45). Baltimore: Johns Hopkins University Press.

Rehmann-Sutter, C. (1996). Frankensteinian knowledge? *Monist, 79,* 264–279.

Roe, S. A. (1981). *Matter, life, and generation.* Cambridge: Cambridge University Press.

Rose, S. P. R. (1981). From causations to translations: What biochemists can contribute to the study of behavior. In P. P. G. Bateson & P. H. Klopfer (Eds.), *Perspectives in ethology* (Vol. 4) (157–177). New York: Plenum.

Rosenblatt, J. S., & Siegel, H. I. (1981). Factors governing the onset and maintenance of maternal behavior among nonprimate mammals. In D. J. Gubernick & P. H. Klopfer (Eds.), *Parental care in mammals* (13–76). New York: Plenum.

Rosenthal, P. (1984). *Words and values: Some leading words and where they lead us.* Oxford: Oxford University Press.

Ryle, G. (1949). *The concept of mind.* New York: Harper and Row.

Sahlins, M. D. (1976). *The use and abuse of biology.* Ann Arbor: University of Michigan Press.

Sarkar, S. (1996). Biological information: A skeptical look at some central dogmas of molecular biology. In S. Sarkar (Ed.), *The philosophy and history of molecular biology: New perspectives* (187–231). Dordrecht, Netherlands: Kluwer Academic.

Schaffner, K. F. (1998). Genes, behavior, and developmental emergentism: One process, indivisible? *Philosophy of Science 65,* 209–252.

Schatten, G., & Schatten, H. (September/October 1983). The energetic egg. *Sciences,* 28–34.

Schneirla, T. C. (1966). Behavioral development and comparative psychology. *Quarterly Review of Biology, 41,* 283–302.

Schrödinger, E. (1967). *What is life?* and *Mind and matter.* Cambridge: Cambridge University Press.

Searle, J. R. (1992). *The rediscovery of the mind.* Cambridge: MIT Press/Bradford Books.

Shedletsky, L. (1980). Can we use our brains to define communication? *Encoder, 8,* 30–40.

Shklar, J. N. (1971). Subversive genealogies. In C. Geertz (Ed.), *Myth, symbol, and culture* (129–155). New York: Norton.

Shotter, J. (1993). Bakhtin and Vygotsky: Internalization as a boundary phenomenon. *New Ideas in Psychology, 3,* 379–390.

Sidanius, J., Cling, B. J., & Pratto, F. (1991). Ranking and linking as a function of sex and gender role attitudes. *Journal of Social Issues, 47* (3), 131–149.

Smith, B. C. (1996). *On the origin of objects.* Cambridge: MIT Press.

Smith, B. H. (1997). *Belief and resistance: Dynamics of contemporary intellectual controversy.* Cambridge: Harvard University Press.

Smith, B. H. (February 20, 1998). Is it really a computer? [Review of *How the mind works*]. *Times Literary Supplement,* 3–4.

Smith, K. C. (in press). What is a genetic trait? In D. Magnus (Ed.), *Contemporary genetic technology: Ethical, legal and social challenges.* Melbourne, Fla.: Krieger Publishing.

Sober, E. (1983). Mentalism and behaviorism in contemporary psychology. In D. W. Rajecki (Ed.), *Comparing behavior: Studying man studying animals* (113–142). Hillsdale, N.J.: Erlbaum.

Sober, E. (Ed.) (1984). *Conceptual issues in evolutionary biology.* Cambridge: MIT Press.

Sommerhoff, G. (1974). *Logic of the living brain.* New York: Wiley.

Stebbins, G. L. (1972). The evolutionary significance of biological templates. In A. D. Breck & W. Yourgrau (Eds.), *Biology, history, and natural philosophy* (79–102). New York: Plenum.

Stent, G. S. (Ed.). (1980). *Morality as a biological phenomenon.* Berkeley: University of California Press.

Stent, G. S. (1986). Hermeneutics and the analysis of complex biological systems. In D. J. Depew & B. H. Weber (Eds.), *Evolution at a crossroads* (209–225). Cambridge: MIT Press.

Sterelny, K. (in press). Development, evolution and adaptation. *Philosophy of Science.*

Sterelny, K., Smith, K. C., & Dickison, M. (1996). The extended replicator. *Biology and Philosophy, 11,* 377–403.

Stich, S. P. (1975). Introduction. In S. P. Stich (Ed.), *Innate ideas* (1–22). Berkeley: University of California Press.

Strohman, R. C. (March 1997). The coming Kuhnian revolution in biology. *Nature Biotechnology, 15,* 194–200.

Symons, D. (1979). *The evolution of human sexuality.* Oxford: Oxford University Press.

Symons, D. (1987). If we're all Darwinians, what's the fuss about? In C. B. Crawford, M. S. Smith, & D. Krebs (Eds.), *Sociobiology and psychology* (121–146). Hillsdale, N.J.: Erlbaum.

Taylor, P. J. (1995). Building on construction: An exploration of heterogeneous constructionism, using an analogy from psychology and a sketch from socio-economic modeling, *Perspectives on Science 3* (1) 66–98.

Taylor, P. J., & García-Barrios, R. (1995). The social analysis of ecological change: From systems to intersecting processes. *Social Science Information 34,* 5–30.

Taylor, R. (1967). Causation. In P. Edwards (Ed.), *The encyclopedia of philosophy* (Vol. 2) (56–66). New York: Macmillan and Free Press.

Thelen, E., & Smith, L. B. (1994). *A dynamic systems approach to the development of cognition and action.* Cambridge: MIT Press/Bradford Books.

Thomas, A., & Chess, S. (1977). *Temperament and development.* New York: Brunner/Mazel.

Thompson, J. N. (1982). *Interaction and coevolution.* New York: Wiley.

Tinbergen, N. (1963). On aims and methods in ethology. *Zeitschrift für Tierpsychologie, 20,* 410–433.

Tobach, E. (1972). The meaning of the cryptanthroparion. In L. Ehrman, G. S. Omenn, & E. Caspari (Eds.), *Genetics, environment and behavior* (219–239). New York: Academic Press.

Tobach, E., & Greenberg, G. (1984). The significance of T. C. Schneirla's contribution to the concept of levels of integration. In G. Greenberg & E. Tobach (Eds.), *Behavioral evolution and integrative levels* (1–7). Hillsdale, N.J.: Erlbaum.

Tomback, D. F. (1983). Nutcrackers and pines: Coevolution or coadaptation? In M. H. Nitecki (Ed.), *Coevolution* (179–224). Chicago: University of Chicago Press.

Toulmin, S. (1967). Neuroscience and human understanding. In G. C. Quarton, T. Melnechuk, & F. O. Schmitt (Eds.), *The neurosciences: A study program* (822–832). New York: Rockefeller University Press.

Toulmin, S. (1970). Reasons and causes. In R. Borger & F. Cioffi (Eds.), *Explanation in the behavioural sciences* (1–26). Cambridge: Cambridge University Press.

Toulmin, S. (Ed.) (1982). *The return to cosmology.* Berkeley: University of California Press.

Toulmin, S., & Goodfield, J. (1962). *The architecture of matter.* Chicago: University of Chicago Press.

Turkewitz, G., & Devenny, D. A. (Eds.). (1993). *Developmental time and timing* (125–142). Hillsdale, N.J.: Erlbaum.

van der Weele, C. (1999). *Images of development: Environmental causes in ontogeny.* Albany, N.Y.: State University of New York Press.

van Gelder, T. (1995). What might cognition be, if not computation? *Journal of Philosophy, 91,* 345–381.

Van Valen, L. (1983). How pervasive is coevolution? In M. H. Nitecki (Ed.), *Coevolution* (1–19). Chicago: University of Chicago Press.

Varela, F. J., Thompson, E., & Rosch, E. (1991). *The embodied mind.* Cambridge: MIT Press.

Vayda, A. P. (1996). *Methods and explanations in the study of human actions and their environmental effects.* Jakarta, Indonesia: CIFOR/WWF.

Waal, F. de (1996). *Good natured: The origins of right and wrong in humans and other animals.* Cambridge: Harvard University Press.

Waddington, C. H. (1957). *The strategy of the genes.* London: George Allen and Unwin.

Waddington, C. H. (1960). *The ethical animal.* Chicago: University of Chicago Press.

Waddington, C. H. (1962). *New patterns in genetics and development.* New York: Columbia University Press.

Waddington, C. H. (Ed.) (1972). *Towards a theoretical biology* (Vol. 4). Edinburgh: Edinburgh University Press.

Waddington, C. H. (Ed.) (1975). *The evolution of an evolutionist.* Ithaca, N.Y.: Cornell University Press.

Weber, B. H., & Depew, D. J. (1999). Does the second law of thermodynamics refute the neoDarwinian synthesis? In P. Koslowski (Ed.), *Sociobiology and bioeconomics. The theory of evolution in biological and economic theory* (50–75). Studies in Economic Ethics and Philosophy (Vol. 20). Berlin: Springer-Verlag.

Webster, G., & Goodwin, B. C. (1982). The origin of species: A structuralist approach. *Journal of Social and Biological Structures, 5,* 15–47.

Weil, A. (1972). *The natural mind.* Boston: Houghton Mifflin.

Weiss, P. A. (1966). Specificity in the neurosciences. In F. O. Schmitt & T. Melnechuk (Eds.), *Neurosciences research symposium summaries* (179–212). Cambridge: MIT Press.

Weiss, P. [A.] (1967). $1 + 1 \neq 2$ (One plus one does not equal two). In G. C. Quarton, T. Melnechuk, & F. O. Schmitt (Eds.), *The neurosciences: A study program* (801–821). New York: Rockefeller University Press.

Weiss, P. A. (1969). The living system: Determinism stratified. In A. Koestler & J. R. Smythies (Eds.), *Beyond reductionism* (3–55). New York: Macmillan.

Weiss, P. A. (1978). Causality: linear or systematic? In G. A. Miller and E. Lenneberg (Eds.), *Psychology and biology of language and thought: Essays in honor of Eric Lenneberg* (13–26). New York: Academic Press.

Weisstein, N. (1971). Psychology constructs the female. In V. Gornick & B. K.

Moran (Eds.), *Woman in sexist society* (207–224). New York: New American Library (Signet).

White, H. V. (1972). What is a historical system? In A. D. Breck & W. Yourgrau (Eds.), *Biology, history, and natural philosophy* (233–242). New York: Plenum.

Whyte, L. L. (1965). Internal factors in evolution. New York: Braziller.

Wicken, J. S. (1979). Entropy and evolution: A philosophic review. *Perspectives in Biology and Medicine, 22,* 285–300.

Wiener, N. (1967). *The human use of human beings.* New York: Avon Books.

Williams, G. C. (1974). *Adaptation and natural selection.* Princeton: Princeton University Press.

Wilson, E. O. (1975). *Sociobiology: The new synthesis.* Cambridge: Harvard University Press/Belknap Press.

Wilson, E. O. (1978). *On human nature.* Cambridge: Harvard University Press.

Wimsatt, W. C. (1986). Developmental constraints, generative entrenchment, and the innate-acquired distinction. In W. Bechtel (Ed.), *Integrating scientific disciplines* (185–208). Dordrecht, Netherlands: Martinus Nijhoff.

Wimsatt, W. C. (1999). Generativity, entrenchment, evolution, and innateness. In V. Hardcastle (Ed.), *Biology meets psychology: Constraints, connections, conjectures* (139–179). Cambridge: MIT Press.

Yund, M. A., & Germeraad, S. (1980). *Drosophila* development. In T. Leighton & W. F. Loomis (Eds.), *The molecular genetics of development* (237–360). New York: Academic Press.

Zolman, J. F. (1982). Ontogeny of learning. In P. P. G. Bateson & P. H. Klopfer (Eds.), *Perspectives in ethology* (Vol. 5) (275–323). New York: Plenum.

Index of Names

For endnotes, page indicates where name appears, not where note begins.

DeFries, J. C., 35, 180–181
DeMaeyer, L. C. M., 74
Dennett, D. C., 163, 195–198, 200, 201, 207, 212–214
Dentan, R. K., 96–97
Depew, D. J., 211
Descartes, R., 14, 62, 73, 124, 184, 230 n22
Devenny, D. A., 206
Dickison, M., 207
Dinnerstein, D., 23, 108
Dixon, R. A., 22
Doolittle, W. F., 17
Doyle, R. M., 242 n14
Dretske, F. I., 225 n10
Driesch, H., 15
Dunbar, R. I. M., 48

Eibl-Eibesfeldt, I., 2
Eigen, M., 45, 74
Elman, J. L., 212
Elsdale, T., 133–134, 183, 233 n2, n3

Fausto-Sterling, A., 240 n5
Fernberger, S. W., 118–121
Fishbein, H. D., 95, 107, 109–116, 120, 227 n15, 228 n17
Fodor, J., 163, 224 n6
Ford, D. H., 241 n13
Foster, D. B., 75, 232 n1
Fredrick, J. F., 145
Freud, S., 108, 112

Galanter, E., 62, 64, 71, 78
García-Barrios, R., 202
Gasking, E., 217 n1, n2, 218 n3, 220 n2, 221 n5
Gatlin, L. L., 31, 69, 74, 76, 77, 81
Geertz, C., 165, 179
Germeraad, S., 36
Gesell, A., 31, 34–35, 37, 99
Gibson, J. J., 241 n13
Glazier, S. C., 222 n2 (ch. 4)

Godfrey-Smith, P., 241 n12, n13
Goldberg, R., 223 n4
Goodfield, J., 14
Goodwin, B. C., 57–58, 69, 134, 155, 234 n4
Gordon, D. M., 208
Gottlieb, G., 47, 153, 163, 208, 240 n7, n8, 241 n11
Gould, S. J., 29–30, 175, 218 n3, 220 n3, 221 n1
Gray, R. D., 207, 211, 239 n4, 241 n10, n11, n12
Greenberg, G., 174, 218 n4
Griesemer, J., 211, 241 n10
Griffin, D. R., 230 n22
Griffiths, P. E., 207, 211, 241 n12
Grmek, M. D., 73, 174
Grobstein, C., 15
Gruber, H. E., 183–184, 218 n3, 237 n6

Haila, Y., 241 n11
Hailman, J. P., 14, 79, 137
Haller, A. von, 171–172
Hamburger, V., 160–176
Harris, M., 120, 228 n19, 231 n23
Heider, F., 22
Hendriks-Jansen, H., 242 n15
Hershey, A., vii
Hinde, R. A., 15, 54, 76, 219 n2, 229 n20, 237 n5
Ho, M.-W., 135, 236 n4
Hobbes, T., x
Hofer, M. A., 142
Hoffman, R. J., 152
Hofstadter, D. R., 23–24, 71, 79, 222–223 n4, 224 n6
Hull, D. L., 221 n1
Hultsch, D. F., 22
Hyland, M. E., 7, 219 n4

Immelmann, K., 167
Ingold, T., 239 n4, 242 n15

Jablonka, E., 208, 240 n7
Jacob, F., 12, 16, 31, 35, 46, 74, 76,
  77, 85, 99, 217 n2, 218 n3, 224 n7,
  225 n9
Jacobson, M., 85
Jerne, N. K., 85
Jervis, R., 202, 240 n6
Johnston, T. D., 63, 152–253, 155,
  208, 235 n6, 240 n8, 241 n11, n13
Jusczyk, P. W., 236 n4

Kagan, J., 225 n2, 229 n20
Kamin, L. J., 218 n4
Kant, I., 13, 173, 184, 234 n4
Kaufman, T. C., 24, 67, 145–147,
  175, 176, 183, 212, 224 n4
Keegan, R. T., 183
Keller, E. F., 238 n1 (Afterword)
King, J. A., 181
Kitchener, R. F., 6
Kitcher, P., 241 n12
Klama, J., 240 n7
Klopfer, P. H., 15, 76, 89, 152, 162,
  172, 184
Knight, R. D., 241 n12
Koestler, A., 220 n4
Kolata, G., 38
Konner, M., 95–101, 107, 125, 226
  n9, 227 n14
Kroeber, A., 219 n2
Kummer, H., 111–112

Lamarck, J. B., 5, 33–34, 99, 176
Lamb, M. J., 208, 240 n7
Lehrman, D. S., 15, 54, 66, 98, 108,
  230 n22
Leighton, T., 38
Lerner, R. M., 6, 22, 227, 229,
  241 n13
Levins, R., 239 n4
Lewin, R., 223 n4
Lewis, J., 233 n3
Lewontin, R. C., 14, 17, 45, 91, 137,

138, 218 n4, 223 n4, 236 n3, 239
  n4
Lindee, M. S., 238 n1 (Afterword)
Long, C. H., 1
Loomis, W. F., 38
Lorenz, K., 55, 57, 64–67, 85, 109,
  213, 234 n4
Løvtrup, S., 135, 233 n3
Lumsden, C. J., 59–61, 81, 85, 107,
  117, 218 n3, 222 n2
Lysenko, T. D., 5

MacWilliams, H. K., 134
Margulis, L., 36, 142, 145
Martin, P., 240 n7
Mattern, R., 101
Maupertuis, P. L., 217 n2, 221 n5
May, R. M., 46
Mayr, E., 30, 61–64, 149, 222–223
  n4
McClearn, G. E., 35, 180–181
McFarland, J. D., 234 n4
Mead, G. H., 22
Mehler, J., 236 n4
Mendel, G., ix
Mercer, E. H., 156
Merchant, C., 87
Midgley, M., 89, 95, 102–106, 141,
  227 n13
Miller, G. A., 62, 64, 71, 78, 133
Monod, J., 31–34, 37, 99, 197,
  224 n7
Montague, M. F. A., 50
Montalenti, G., 14, 31
More, H., 73, 184
Morton, J., 236 n4
Moss, L., 242 n14
Munn, N. L., 118–121
Murphy, C., 195

Nagel, E., 155, 235 n6
Needham, J., 147
Nelkin, D., 238 n1 (Afterword)

Index of Subjects

as progressive, 33, 46, 87, 172, 183
as replacement for God, 31, 62, 89
as succession of developmental systems, 39, 117, 138, 147. *See also* developmental systems, evolution of
as tinkering, 46, 153

fate, 8–9, 37, 64, 87, 137, 158, 161, 164, 186, 238 n1
feedback, 38–39, 78, 131, 135
form, 12, 30, 90, 124, 158
    construction of, 7, 26, 35
    emergence of, 3, 26, 38–39, 133, 226 n5, 234 n5
    imposition of, 1, 14–15, 26, 35, 219 n4, 225 n9
    origin of, ix–x, 1–2, 14, 35, 183
    preexisting, 1, 3, 29, 31, 53, 57, 60, 135, 160
    transmission of, 5, 16, 26, 31, 131
freedom, 83, 87, 90–92, 188, 191–193, 229 n22, 236 n3
function, 62, 149, 197, 198

gene, ix–xv, 88, 191, 204, 214. *See also* DNA
    activation of, 38, 69, 145, 156
    as agent, 1, 8, 63, 74, 114, 124, 160, 173, 195, 197, 201, 210 (*see also* homunculoid gene)
    cognitive-causal, 12, 14, 17, 39, 54, 90, 109, 145, 176–177, 194–195 (*see also* metaphor, cognitive-causal; gene, as agent; homunculoid gene)
    as developmental interactant, 40, 131
    homunculoid. *See* homunculoid gene

as prime cause of development, 13, 30–33, 40, 156, 176, 198
selfish, 140, 157, 176–177, 195, 232 n24
as vehicle of inherited form, 6, 16, 99
genetic blueprints, 12, 15, 54–55, 70
genetic code, 16, 23, 25, 58, 69, 77, 81–82. *See also* code
genetic hypotheses, 56–59
genetic information. *See* information, genetic
genetic instructions. *See* genetic programs
genetic knowledge, 55
genetic message. *See* message, genetic
genetic plans. *See* genetic blueprints
genetic programs, 12, 15, 24, 35–36, 49, 61–74, 205, 206, 230 n22. *See also* programs
    descriptive vs. prescriptive, 63
    as products of evolution, 61, 99, 147, 205–206, 222 n4
genetic symbols, 56–59
genetics, 145, 160, 176
    and behavior, 7, 180–181
genotype, 16–17, 48, 61, 164, 177, 180–181, 209–210, 221 n1
"ghost in the machine," 8, 60, 88, 95, 113, 118, 133, 156. *See also* homunculoid gene; homunculus
    and "man in the head," 12, 14, 84, 88–89, 128, 133, 159 (*see also* homunculoid gene; homunculus)
goal direction, 63, 90, 149–154, 186, 235 n6
goals, 60, 61, 89, 158, 185–187
God, 1, 15, 88–90, 94, 95, 161, 164, 172, 187, 189

Susan Oyama is Professor of Psychology at the John Jay
College of Criminal Justice, CUNY Graduate Center. She
is the author of *Evolution's Eye: A Systems View of the
Biology-Culture Divide* (Duke University Press, 2000).

Library of Congress Cataloging-in-Publication Data

Oyama, Susan.
The ontogeny of information : developmental systems and
evolution. — Rev. and enl. ed. / Susan Oyama ; foreword
by Richard C. Lewontin.
p. cm. — (Science and cultural theory)
Includes bibliographical references and index.
ISBN 0-8223-2431-8 (cloth : alk. paper). —
ISBN 0-8223-2466-0 (pbk. : alk. paper)
1. Information theory in biology.   2. Evolution (Biology)
I. Title.   II. Series.
QH507.O93   2000
576.8—dc21   99-34735